Embodying Bioethics

T0127902

New Feminist Perspectives Series
Series Editor: Rosemarie Tong, Davidson College

Claiming Reality: Phenomenology and Women's Experience
    by Louise Levesque-Lopman
Evidence on Her Own Behalf: Women's Narrative as Theological Voice
    by Elizabeth Say
Feminist Jurisprudence: The Difference Debate
    edited by Leslie Friedman Goldstein
Is Women's Philosophy Possible?
    by Nancy J. Holland
Manhood and Politics: A Feminist Reading of Political Theory
    by Wendy L. Brown
"Nagging" Questions: Feminist Ethics in Everyday Life
    edited by Dana E. Bushnell
Rethinking Ethics in the Midst of Violence: A Feminist Approach to Freedom
    by Linda A. Bell
Speaking from the Heart: A Feminist Perspective on Ethics
    by Rita C. Manning
Take Back the Light: A Feminist Reclaimation of Spirituality and Religion
    by Sheila Ruth
Toward a Feminist Epistemology
    by Jane Duran
Voluptuous Yearnings: A Feminist Theory of the Obscene
    by Mary Caputi
Women, Militarism, and War: Essays in History, Politics, and Social Theory
    edited by Jean Bethke Elshtain and Sheila Tobias
Women, Sex, and the Law
    by Rosemarie Tong
Women and Spirituality, Second Edition
    by Carol Ochs
Rethinking Masculinity, Second Edition
    by Larry May, Robert Strikwerda, and Patrick D. Hopkins
Pandora's Box: Feminism Confronts Reproductive Technology
    by Nancy Lublin
*Analyzing the Different Voice: Feminist Psychological Theory
and Literary Texts*
    edited by Jerilyn Fisher and Ellen S. Silber

Embodying Bioethics

Recent Feminist Advances

edited by
ANNE DONCHIN
and
LAURA M. PURDY

ROWMAN & LITTLEFIELD PUBLISHERS, INC.
Lanham • Boulder • New York • Oxford

ROWMAN & LITTLEFIELD PUBLISHERS, INC.

Published in the United States of America
by Rowman & Littlefield Publishers, Inc.
4720 Boston Way, Lanham, Maryland 20706

12 Hid's Copse Road
Cumnor Hill, Oxford OX2 9JJ, England

Copyright © 1999 by Rowman & Littlefield Publishers, Inc.

All rights reserved. No part of this publication may be reproduced,
stored in a retrieval system, or transmitted in any form or by any
means, electronic, mechanical, photocopying, recording, or otherwise,
without the prior permission of the publisher.

British Library Cataloguing in Publication Information Available

Library of Congress Cataloging-in-Publication Data

Embodying bioethics : recent feminist advances / edited by Anne
    Donchin and Laura M. Purdy.
        p.   cm.
    Includes bibliographical references and index.
    ISBN 0-8476-8924-7 (cloth : alk. paper). — ISBN
0-8476-8925-5 (pbk. : alk. paper)
    1. Bioethics.   2. Feminist theory.   I. Donchin, Anne.   II. Purdy,
Laura Martha.   III. International Association of Bioethics.
QH332.E43   1999
174'.dc21   98-28146
                                                                  CIP

Printed in the United States of America

∞ ™ The paper used in this publication meets the minimum requirements of
American National Standard for Information Sciences—Permanence of Paper
for Printed Library Materials, ANSI Z39.48–1984.

Contents

Preface   vii

    Introduction   1
    Anne Donchin and Laura M. Purdy

Part I: Redirecting Bioethical Theory

1 Rehabilitating Care   17
    Alisa L. Carse and Hilde Lindemann Nelson

2 Just Caring About Maternal–Fetal Relations: The Case of
    Cocaine-Using Pregnant Women   33
    Rosemarie Tong

3 Closing the Gaps: An Imperative for Feminist Bioethics   45
    Helen Bequaert Holmes

4 Erasing Difference: Race, Ethnicity, and Gender in
    Bioethics   65
    Susan M. Wolf

Part II: Reproduction and Beyond

5 Abortion, Chernobyl, and Unanswered Genetic Questions   85
    Laura Shanner

6 Should Lesbians Count as Infertile Couples? Antilesbian
    Discrimination in Assisted Reproduction   103
    Julien S. Murphy

7 Equality, Autonomy, and Feminist Bioethics   121
    Elisabeth Boetzkes

8   Health Commodification and the Body Politic: The
    Example of Female Infertility in Modern China   141
    Lisa Handwerker

9   Feminism and Elective Fetal Reduction   159
    Mary V. Rorty

10 On Not Iterating Women's Disability: A Crossover
    Perspective on Genetic Dilemmas   177
    Anita Silvers

11 Menopause: Is This a Disease and Should We Treat It?   203
    Wendy A. Rogers

Part III: Working for Change

12 Culture and Reproductive Health: Challenges for Feminist
    Philanthropy   223
    Nikki Jones

13 Strategies for Effective Transformation   239
    Barbara Nicholas

14 Women and Health Research: From Theory, to Practice,
    to Policy   253
    Françoise Baylis, Jocelyn Downie, and Susan Sherwin

Index   269

About the Editors and Contributors   283

Preface

This volume brings under one cover the labors of feminist scholars from diverse backgrounds. It would not have come into being were it not for the International Network on Feminist Approaches to Bioethics (FAB). Jointly conceived by Anne Donchin and Helen Bequaert (Becky) Holmes and brought forth at the Inaugural Congress of the International Association of Bioethics in 1992, FAB has matured at a swift pace without compromising the mission proclaimed in its initial call for members: "development of a more inclusive theory of bioethics encompassing the standpoints and experiences of women and other marginalized social groups, reexamination of the principles and legitimizing functions of the prevailing discourse, and creation of new strategies and methodologies."

In its brief life, FAB has acquired a membership that includes feminist bioethicists from virtually all the major English-speaking countries as well as Argentina, Austria, Belgium, Brazil, Germany, India, Italy, Japan, the Netherlands, the Philippines, and Spain. Though many of us have never met face-to-face, our listserv—established in 1993 and still maintained by Corinne Bekker at the University of Utrecht—expedites cyberspace conversation. For those who lack e-mail access or who still prefer the look and feel of a hard copy, there is FAB's semiannual newsletter, now in its sixth year of publication. Over the course of the newsletter's initial four years, Becky Holmes valiantly struggled through the production process (with intermittent help from a few stalwarts). Then in 1997, through the intervention of Margaret Little, it finally acquired much needed institutional support at the Kennedy Institute of Ethics. FAB continues to maintain its original connection with the IAB, and one of our own, Susan Sherwin, is now in her second term on its board. In 1996, in conjunction with its Third World Congress in San Francisco, FAB organized its own conference, attended by over one hundred participants, and in November 1998 a successor conference will be held in Japan. The essays included in this volume developed mostly from presentations at the San Francisco conference.

One of us (Anne Donchin) chaired the group that organized the initial FAB conference and sought out a publisher. In keeping with FAB's commitment to an egalitarian structure, a call was announced on the listserv inviting nominations for coeditor. Longstanding FAB member and supporter Laura Purdy confesses that her first reaction was, "Don't even think about it!" But, she later confessed: "The project crouched like a cat at the back of my mind." Concurrent sessions, chats with friends and colleagues, and the limits of the human mind (not to mention the human rump) meant we would inevitably miss a lot of papers in San Francisco. The anthology would give us an opportunity to savor them later, and our work would provide a forum for creative new voices in feminist bioethics.

The introduction that follows, like our work reviewing and editing the essays included here, has been a collaborative labor. Like good contemporary domestic partners, we took turns pulling on the oars as circumstances required. When either of us was preoccupied with other chores, the other grasped the initiative. While Anne was keeping tabs on potential contributors during the summer lull, Laura, secluded in the Greek countryside with no phone and only rudimentary mail service, took to reflecting on recent developments in feminist bioethics. The initial section of our introduction distills her musings under the shade of an almond tree. Then, as we got down to the final group of essays, Anne started making interconnections among them. So the sections that follow that focus on the book's contents are largely hers. Both of us regret the limitations of space that forced us to leave out worthy potential contributions. If this collection establishes a FAB tradition, however, you'll meet them in a second volume.

We are particularly grateful to all who contributed to the spectacular success of the first Feminist Approaches to Bioethics conference in San Francisco: to Joan Callahan, Mary Mahowald, Laura Shanner, Barbara Nicholas, and Susan Sherwin, who took on the responsibility of gathering like-minded scholars together and arranging topical sessions; to Alex Capron and his staff at the Pacific Center for Health Policy and Ethics, who shouldered mountains of organizational detail and still managed to house us all in queenly style; to Jacquelyn Ann Kegley and Rosalind Ekman Ladd, who gave of themselves generously to trying tasks; and especially to Helen Bequaert Holmes, who lent indispensable support and expertise at every stage in the development of this project. Thanks, too, to Rosemarie Tong for her commitment to the FAB

Network and her faith that our fledgling project would come to fruition. And finally, thanks to Jennifer Purvis, Patrycja Maksalon, and Maya Goldenberg, without whose assistance the manuscript would not have come in anywhere near its deadline.

Anne Donchin
Laura Purdy

Introduction

Anne Donchin and Laura M. Purdy

Context: A Decade in the Young Life of Feminist Bioethics

The contributions in this volume raise fundamental and critical questions about feminism's encounter with bioethics that could not have been anticipated even a decade ago. There is now more thinking, writing, publishing, teaching, and organizing in feminist bioethics than ever before. It was hard enough to keep up in 1988, when Becky Holmes and Laura Purdy set to work editing the two issues of *Hypatia* on feminist bioethics. Now it is almost impossible. On the less bright side, politics in the United States (and in many other societies) have continued to favor policies benefiting privileged segments of society at the expense of marginalized groups, such as women generally, ethnic and racial minorities, persons with disabilities, and gays and lesbians. Some of the worst damage has been done to health and health care. On the one hand, factors affecting human health (sanitation, clean food and water, decent housing, occupational safety, environmental protection, and education) continue to be eroded by cuts in social programs. On the other, 41 million Americans now lack health insurance (up from 36 million in the mid-1990s), and millions more are inadequately insured; the public hospitals that used to deliver "charity" care to people in these circumstances are facing budget cuts or are being closed altogether. Much of the rest of the population is being encouraged or forced into for-profit managed care plans and health maintenance organizations. The previous tendency toward overtreatment of those with good insurance has abated somewhat, to be replaced, far too often, with new barriers to care for those who do develop medical problems.

Editing projects such as this one provoke rethinking of the intersection between feminist theorizing and the practical realities that afflict

1

contemporary health care. Both the erosion of social programs and changes in the health care system mean more needlessly sick, disabled, or prematurely dying people. Women, because of their reproductive and nurturing roles and their disproportionate poverty and representation among the aged, are bearing more than their share of the burden. One sometimes wonders whether working on the fine points of feminist bioethics isn't fiddling while Rome is burning. Also, academic and practical bioethics have grown enormously in the last few years, but how much of this is informed by feminist concerns? Consider, for example, the 1996 IAB conference, where few prominent bioethicists stayed to hear what we had to say. We clearly need to focus on ways to maximize the impact of our work. Encouraging students and colleagues to take feminism seriously, developing and maintaining high standards, and careful strategizing are all necessary.

A few years ago, Laura pondered the marginalized position of feminism in society at large and in bioethics in particular, and set out to show that justice requires a feminist outlook.[1] She concluded that to incorporate this we need to emphasize basic feminist concerns. Given the much vaunted diversity of feminism these days, this seems like a wildly implausible project. Yet at the heart of most feminist work is a set of judgments that seems central. As Susan Sherwin has pointed out, there are some core views that transcend the differences dividing feminists in their internal debates:

> a recognition that women are in a subordinate position in society, that oppression is a form of injustice and hence is intolerable, that there are further forms of oppression in addition to gender oppression (and that there are women victimized by each of these forms of oppression), that it is possible to change society in ways that could eliminate oppression, and that it is a goal of feminism to pursue the changes necessary to accomplish this.[2]

This position will not satisfy all those who identify themselves as feminists. Postmodernists, for example, doubt the usefulness both of reason and of the category "woman." At this point, however, neither doubt seems compelling. Showing that some version of reason is flawed does not justify rejecting such a crucial concept altogether. "Woman," too, may need some work to ensure that it recognizes morally relevant differences among us, but given its importance as a moral category, it remains to be shown that eradication rather than redefinition is the best approach.

Why is accepting core feminism a good idea? The reasons are practical, political, and philosophical. We live in a confusing era, when some feminist assumptions have been widely internalized, at least by girls and women. Yet "feminism" has been turned into a dirty word by unsympathetic media and equated with manhating, bra burning, selfishness, and "perversion." Worse yet, feminism is (along with other movements for greater equality) a target of immensely powerful forces that want white middle-class women at home, blacks and other minorities "in their place," gays and lesbians in the closet. We can probably not do much about "feminism's" tattered state, for any new word we come up with would be subjected to the same drubbing. We could use the word less, argue for positions without labeling them "feminist." But most of us have paid a high price for identifying ourselves as feminists, and giving up our badge would be painful. In a few cases, it might be a sensible course. Better overall, perhaps, to hold on to our main strategy and continue to emphasize the simple message that justice requires eradicating inequality. Core feminism is committed to that.

Emphasizing core feminism should also help us connect with other justice movements. Such coalition building appears to be the only hope for a future world that makes human welfare its first priority. Progressives sometimes indulge in bitter fights about relatively minor differences of opinion; unfortunately, that balkanizes us at times when a united front might achieve the critical mass it takes to influence policy. Sticking to core feminism would also clarify what is going on as we attempt to reconstruct theory and practice. Sadly, it sometimes looks as if we're merely "packaging" feminism. Feminists are sometimes stereotyped as persons who doubt certain kinds of scientific claims to objectivity, or who believe that morality should promote relationships, or who oppose contract pregnancy. To endorse such stereotypes is to stifle critical thinking and make feminism appear to be just another philosophical or political "ism" that can be taken or left. But something much more important and interesting is actually going on: we are seeking to remake philosophy. Any progress in philosophy should, of course, be adopted by feminists. However—and this is the vital issue— progress made by feminists should be adopted by mainstream philosophers, too. Not because it is feminist, but because it is better. Take, for instance, the oft-repeated claim that dichotomies are unfeminist. You should be wary of them, however, even if you aren't a feminist because dichotomies tend to oversimplify, to narrow vision, and to have subtle and often harmful social consequences. Consequently, the burden of proof for showing that these accusations do not apply in any given case must be on those who use them. Failing to recognize well-founded claims of this sort is bad philosophy, not just bad feminism.

Core feminism seeks to find and eradicate traces of sexism and other oppressions wherever they may be found; it requires us to take nothing for granted, to question everything. This is an exciting enterprise, but also a daunting one. We sit, master's tools in hand, eyeing the master's house. Both tools and house need to be re-formed. To do so, we need to feel free to explore, experiment, try many paths. A good deal of progress has been made both on the tools and on the house itself. But we are nowhere near done, and that's hardly surprising. Western civilization has been in the making thousands of years, and we have been at work for just a few. Not every path we consider will prove fruitful, nor should we expect it to do so. Reconstructing thought, like writing, requires us to start with a rough approximation of our goal, without excessive concern for elegance or precision of the finished product. Clarification, sharpening, revision can come later; and, if what we produce isn't quite right, we must have the courage to change it. Such thinking and rethinking is both an individual and a social task. For ideas will likely grow better if they are subjected to scrutiny by sympathetic but careful onlookers. Those who are further removed can often see weak points that are invisible to us, can often help us find ways to strengthen our arguments or deal with objections.

This process can be wonderfully beneficial, even if sometimes painful. It can also backfire, of course. Critics might dismiss our project altogether or fail to offer constructive help. Their stinging comments may stick in our minds forever. (Some of the nastiest we still remember: "This isn't a philosophy paper at all!" "Why are you writing this paper?" "Whose side are you on, anyway?") We must have the wisdom to discern whether such criticisms arise from fear or from lack of understanding, but also the courage to recognize that there might be some fundamental problem with what we are doing. New insights, whether our own or others', need equally critical scrutiny before we incorporate them into our worldview. Otherwise, our minds will become cluttered with indefensible beliefs, and we will prematurely cease searching for defensible ones. Debate about them needs to be genuinely open.

Naturally, there often will be disagreement about when a new idea ought to be accepted. Core feminism must surely remain open to new ideas, while leaving room to disagree about when to accept them. Having room for disagreement lessens the risk of silencing the cautious with accusations of being disloyal, or not "real feminists." Core feminism is also more open to difference; adopting it should help keep falsely universalizing tendencies at bay. No particular theory is likely to become a necessary part of feminism, blocking recognition of its limited application. The downside is, of course, that no belief, no matter how popular, no matter how deeply held, can become a litmus test, either. So, for example, although most of us are convinced that women will never take their rightful place in society without reproductive choice, and we will argue ourselves hoarse with anybody who believes otherwise, it is possible to deny that position without forsaking feminism. Issues of this kind involve a wide variety of judgments that we can and should continue to debate, such as the moral status of the fetus, but it doesn't make good sense to decree a set of necessary and sufficient conditions that need to be satisfied if a position is to be authentically feminist.

Ticklish questions raised by editing projects such as this one have prompted us to rethink these issues. How should papers be chosen? How much control should be exerted over their content? Their form? What if a paper is original but ultimately seems to argue for an untenable thesis? What if the argument is good but the premises implausible? To what extent should "scholarly" standards be enforced? Such questions seem especially pressing where one wants to ensure that the resulting work is perceived as strong—especially by scholars predisposed to dismiss feminist work as weak. Yet it is also crucial to welcome new ideas and new voices, not stifle them, as can easily happen where senior people control the expression of work by junior people. Morally preferable too, to turn what is otherwise a competitive and stressful experience—all too common in the scholarly world—into a humane one.

In short, providing constructive feedback instead of either a curt "no!" or a scathing attack improves the academic environment. It is especially helpful for writers who may have suffered from lack of mentoring in graduate school and who may now be isolated, under pressure to publish, with few avenues for getting useful feedback. Editing, we firmly believe, should be an extension of feminist pedagogy. We see feminist pedagogy as simply good pedagogy requiring development of our capacities to seek understanding of others' positions and their reasons for believing them. This calls for empathetic connection and sound judgment.

These are high standards, sometimes foundered by anger, despair, or exhaustion. So why not just leave others to do their own moral education? Because teaching often works better! The consequences of a cold shoulder or an angry retort are not likely to increase comprehension or empathy. If people haven't had our experiences, if they really don't understand the issues, if they genuinely want a chance to engage in dialogue, teaching is necessary.

The other side of the coin here is that feminists have a lot to learn from one another and from members of other oppressed groups. In particular, relatively privileged feminists need, and may want, to be taught by those experiencing the manifold other types of discrimination common in our societies. The foregoing considerations apply to them. Just as it cannot be expected that nonfeminists will understand feminism without teaching, white or heterosexual feminists cannot necessarily be expected to understand racism or heterosexism without teaching. Just as anger, impatience, or contempt will not persuade non-feminists, they will not persuade white or heterosexual feminists, either. Just as nonfeminists cannot be expected to know where to turn for information about the oppression of women, white or heterosexual or able-bodied feminists cannot be expected to know where to turn for information about racism or heterosexism or disability discrimination.

These realities do not mean that we are enemies. The real enemies are those who ignore oppression, happily benefit from it, or seek to increase it. The sort of feminist metaethics we propose here could help us all join together in the fight for a more just world. There will, no doubt, be many disagreements, but we hope that they will be more fruitful if discussion is accurate, careful, patient, and kind.

Contents: Toward the Next Decade of Feminist Bioethics

Bioethics Anthologies: Then and Now

The essays in this collection all address facets of the above mentioned realities in their quest to advance feminist bioethics, shifting the direction of bioethical theory and practice from its preoccupation with abstract undifferentiated individuals to the concrete particularities that shape the lives of embodied, socially situated humans. Casting about for ideas to sort the contributions into neat categories presented problems. Anne turned again to Gorovitz et al.'s *Moral Problems in Medicine* (1976), one of the very first bioethics texts.[3] She had used it herself in 1978, the first time she taught bioethics. Now she was struck by two central facts that had initially eluded her.

First, though five of the seven editors were philosophers (including two loyal FAB members, Susan Sherwin and Andrew Jameton) and three of the seven were women (an enviable proportion even today!), only one woman philosopher was included among contributors. Pick up any contemporary anthology and you can't help noticing the difference. But, despite the growing number of women bioethicists, both in philosophy and in related disciplines, the format of bioethics texts has changed remarkably little. More conspicuous than subsequent shifts toward diverse patient populations have been pulls in the opposite direction—toward greater emphasis on abstract universal norms and principles. When that anthology first appeared in 1976, the now ubiquitous moral framework subsequently dubbed "principlism" had not yet made its debut. In the Gorovitz text, the only bow to abstract moral norms was contained in three brief selections from Immanuel Kant,

John Stuart Mill, and (surprisingly) Jean-Paul Sartre. Only sparse attention was given to reproductive issues; much more to abuses of social power in the treatment of such vulnerable groups as children, prisoners, the mentally ill, and the "handicapped." If we gauge the "progress" of bioethics by the signposts Susan Wolf points to in her contribution here, we've been circling backward since 1976! Other marks of progress frequently applauded include the increased presence of bioethicists on public policy panels and among medical school faculty, and the proliferation of conferences devoted to bioethical issues. But, as Becky Holmes asks, whose aims are they serving? Has bioethics, a child of the 1960s U.S. civil rights movement, been co-opted into service to the very powers the civil rights movement struggled to dislodge? And as novel technologies become internationalized and the bioethics movement spreads across cultures, are we also exporting the afflictions feminist bioethicists have identified in the United States?

Barbara Nicholas suggests in her contribution that the way bioethics takes root in a culture may depend more on the characteristics of its native soil than on the original climate in which the technologies were first institutionalized. At least this is how the scenario has played out in New Zealand, where a local scandal exposed the entrenched sexism of the medical community and compelled it to accept more rigorous moral standards of medical research. Several other contributors, including Laura Shanner and Lisa Handwerker, focus on the globalization of first world technologies, their extension into developing economies, and problematic moral consequences for the lives of affected people. Hopefully, readers familiar with circumstances surrounding the "founding moments" of bioethics movements (Wolf's expression) in other countries will be stirred to reflect on the particular mix of local elements and transnational influences at work within their own cultural environments.

The other fact that screamed out from the pages of *Moral Problems in Medicine* was the dearth of articles dealing with reproductive issues. Of course, Louise Brown, the first "test-tube baby", had not yet been conceived in 1976. But the women's health movement already had a good head of steam. *Our Bodies, Ourselves* had been in circulation since 1969.[4] And feminist scholars were decrying the erosion of abortion access, presumably secured by *Roe v. Wade*, and lamenting childbirth practices that sacrificed the interests of the birthing woman to the convenience of her obstetrician. But the only abortion articles that found their way into the Gorovitz collection were those by males: John Noonan, Michael Tooley, and H. Tristram Engelhardt Jr. Involuntary sterilization did make it but was tucked away under the category "paternalism." Times have surely changed here! Editors now have a vast treasure trove of feminist literature to select from, and feminist authors are well represented in the reproduction sections of bioethics texts. Unfortunately, though, stereotypes persist. Reproductive issues have come to be looked on as the special expertise of women. So, though feminist analyses pervade the entire gamut of bioethical themes and topics, comparatively few women and fewer feminists make it to the other sections of these texts. Feminist bioethics is often mistakenly assumed to address "women's concerns"—a special ethics for women.

In selecting material to be included here, we've aimed at striking a balance between issues that bear on reproduction, and other themes and topics that also seize the attention of feminist bioethicists. But issues such as reproduction have come to be so complexly interwoven with other themes in feminist writing that we often found no hard-and-fast way to decide where to place a selection. We chose the easy path, distinguishing initially among essays that offer a broad overview of the shortcomings of the dominant bioethical framework, then turning to those that focus on concerns central to securing reproductive self-determination, and next to other life experiences where combinations of external and internal constraints force uneasy accommodations on us. The concluding section, "Working for Change," chronicles struggles to devise strategies to bring feminist bioethical concerns to center stage where they are more apt to effectively influence social policy.

Redirecting Bioethics

As mainline bioethics has matured, it has given little critical attention to the hierarchical rankings that parcel people into more or less arbitrary groupings: sex, race, ethnicity, age, disability, and susceptibility to genetic disease. To refocus the bioethical enterprise, the articles we have included stress two interrelated themes that are dominant in feminist critiques: power and particularity—the powers that divide and marginalize nondominant people, and the particularities of personal lives that resist confinement within externally imposed categories. These themes, we insist, should be fundamental to bioethics, which has for too long ignored differences in power that mark physician–patient and researcher–subject relationships.

This theme underlies Wolf's forceful critique of the tendency among bioethicists to interpret harms suffered by oppressed groups—even the impoverished black men in the Tuskegee syphilis study or inmates in Nazi concentration camps—as harms solely to generic individuals. By failing to come to terms with group harms, the significance of race, ethnicity, and gender has eluded bioethics, and the standpoints of nondominant groups have been neglected. Constructed from the perspective of an elite group that is blinded to its own partiality, bioethical theory has overlooked such key components of moral life as context, partiality, and relational bonds. This critique is extended by Becky Holmes in her call to demarginalize the "medically disenfranchised," a category that extends to all who are subjected to disparate treatment by medical practitioners and researchers. Holmes points to widening gaps within the U.S. health care system: between the affluent, who are becoming richer, and the poor, who are growing increasingly destitute; between ethics committees that purport to protect patients and the vulnerable people whose fates they determine; and between those at the summit of medical hierarchies and those at the bottom, who do their cleaning up. Holmes's article surveys changes in the terrain since the initial appearance of her coedited *Feminist Perspectives* in 1992[5] and reiterates the call to heal both bioethics and medicine, now a more urgent need than ever.

The contribution herein by Alisa Carse and Hilde Lindemann Nelson echoes Wolf's and Holmes's concerns about power imbalances. Particularly attentive to contextual subtleties and relational bonds, they show how a well-wrought care ethic that is responsive to institutionalized injustices can make a significant contribution to bioethics. Its assimilation into theory and practice can serve as corrective to the all too common tendency among health care providers to regard patients only in their generality—as repeatable subjects of generic care—ignoring particularities essential to understanding the situations of sick and vulnerable individuals. Rosemarie Tong's perceptive treatment of the condition of cocaine-using pregnant women carries this care perspective forward. Her reappraisal tests the limitations of theoretical orientations that spring from the position of the privileged and summons us to embrace with our moral vision the life-world of society's most despised and rejected women. Read in conjunction with the Carse and Nelson contribution, Tong's analysis strengthens the case for incorporating into bioethical theory an ethics of care capable of challenging the structures and systems that perpetuate these women's disempowerment.

Women's Bodies: Reproduction and Beyond

Though reproductive issues have been regarded as women's special province, even this area of scholarship has by no means been exhaustively mined. New issues keep surfacing that are scarcely likely to have the mass-market appeal so critical to inclusion in mainline texts. Some of those we've included here use reproductive issues primarily to illustrate deficiencies in current bioethical thinking and medical practice.

Other contributors are troubled about problems peculiar to procreation and maternity, such as access to reproductive techniques. But they, too, touch on more pervasive issues about the limits of physician authority, the conflation of moral and medical values, and boundaries between definitions of sickness and immorality. Elisabeth Boetzkes's contribution raises a further issue about the state's responsibility to control public meanings of maternity that has recently attracted the attention of feminist scholars through the work of Drucilla Cornell and Moira Gatens.[6] Boetzkes considers how subjective meanings of being female depend on public meanings of maternity, and she points to a paradox in political liberalism: whenever liberal states permit such practices as contract motherhood and sex selection of fetuses, practices that symbolically devalue maternity, liberal values like autonomy and political equality are compromised. Her provocative claims about the liberal state's responsibility to protect women from symbolic harms that stereotype them raise vital questions about feminist strategizing within political settings that often refuse to recognize even gender-specific physical harms. Her attention to symbolic harms opens fresh avenues for debate, too, particularly within a climate where thinkers privileged to have the ear of public policy makers insist that, in the absence of foreseeable physical injury and violations of individual rights, there are no moral reasons to hold back on investigation of innovative techniques such as human cloning. Though Sigmund Freud may well have been right to point out that even a cigar may be only a cigar, it's not so easy to imagine a world where a human clone is just another human life!

Issues raised by Anita Silvers in her discussion of problems confronted by disabled women overlap some of the concerns articulated by Boetzkes, particularly insofar as they intersect abortion decisions that are based on fetal disability. Silvers touches on this issue as but one instance of numerous flawed social practices that contribute to the marginalization of disabled persons. By failing to integrate into their agenda the perspectives of disabled women, Silvers argues, feminists have done a disservice to women with physical, sensory, and cognitive impairments that intensify the social disadvantages they must overcome to live productive lives. She challenges feminists to recognize that, like disempowerment of women generally, disability is less a product of immutable physical characteristics than mutable social arrangements.

Julien Murphy's analysis of alternative strategies to extend access to assisted reproduction to lesbian partners sparks questions about political strategies at another level. Like so much feminist writing, her attention to these issues emerges from the juncture of her professional training and her personal experience. Her own encounter with the health care system defeats the common medical presumption that infertility is a problem only for heterosexual married women defined by their relationship to men. Mary Rorty's essay, like Murphy's, challenges physicians' use of their social power to control access to new techniques. Rorty's worries focus around the still too common tendency of physicians to preempt women's decision-making options by tailoring treatment in ways that advance the physician's goals rather than those of the patient. She shows how this tendency plays out where decisions may need to be made about (s)elective termination of fetuses. Her essay shows why physicians need to give more serious attention to planning a course of treatment collaboratively to reduce occasions when such painful decisions are suddenly thrust upon unprepared patients.

Clinicians tend to vacillate between two poles: an unthinking voluntarism that offers patients (such as pregnant women) an undiscriminating "choice" among the assorted techniques in their armamentaria, with little attention to individualizing risks and benefits to the particular patient under treatment or a rigid paternalism that silences those experiencing the symptoms and imposes on them a statistical norm that may have no bearing on their particular situation. Wendy Rogers addresses this latter group of issues in her account of how the medical establishment and pharmaceutical industry have profiteered from the situation of menopausal and postmenopausal women by redescribing normal bodily changes as inferior anomalies. She artfully utilizes the conceptual tool kit of the physician to debunk the pervasive myth that menopause is a disease. Read in conjunction with Rorty's article, it illustrates practitioners' proclivity to conflate moral and medical values, either by an excess of moral zeal that preempts patients' decisions or by a moral laxity that slights responsibility for risky treatments that are unlikely to yield proportional benefit to patients and may actually contribute to social harms (e.g., sex selection techniques).

Two of the essays that concentrate on reproductive issues deal with social harms that befall formerly socialist countries as they shift toward market economies and appropriate morally dubious Western technological practices, particularly tendencies to view women's bodies as (often faulty) reproductive machines. Laura Shanner chronicles the incredulous official response to reports of genetic anomalies in the wake of the Chernobyl disaster, the sparse funding provided by international governments and research institutions to support data collection, and the continuing distrust among scientists in the East and West who employ different research methodologies. Meanwhile, pregnant women potentially affected by the drifting cloud of radioactive particles are left to grapple with stress and anxiety about prospects of genetically mutated offspring. Barbara Nicholas's trenchant observation is apt: in both developed and developing countries, it is assumed that the effect of a practice on women is only an incidental side effect that raises "women's questions," which, of course, can readily be deferred until later.

An analogous situation is springing up in China as the government unwittingly facilitates the growth of an infertility industry at the same time as it enforces policies stressing the urgent need to reduce the birthrate. Baffling as this anomaly seems, it's readily understandable once one realizes how Western scientific practices lend themselves to the purposes of non-Western leaders. By coupling Western technology with social power, they can more efficiently intensify control of women by a medical establishment that utilizes their bodies as reproducers of their own distinctive cultural norms. Lisa Handwerker intends her account to be read as a chronicle of how a cross-cultural perspective on bioethical issues can provoke us to reflect on intersections between specific technologies and the social, political, and economic structures of countries that incorporate them into their distinctive practices.

Working for Change

The concluding three articles all consider a variety of strategies for transforming bioethics and medical practice in ways that will heighten responsiveness to the situation of women and other marginalized groups. Nikki Jones, Barbara Nicholas, and Françoise Baylis and company all examine different strategies for interjecting women's standpoints into policy-making processes. Their contributions provide a fitting complement to the problems of exclusion addressed in the initial section by Becky Holmes and Susan Wolf. Nikki Jones addresses strategies she has developed to effect change within developing countries, schemes that empower people to address problems they have identified in terms of their own cultural beliefs and interests. She offers refreshing examples of how a first world donor agency can put its power and influence to work to reduce maternal death and morbidity without replicating power relationships under colonialism. For instance, a cleverly devised system of grants to indigenous women's organizations actually induced male-led groups to collaborate to implement reforms in marriage and childbirth practices. Her essay demonstrates why women within countries where harmful practices such as genital surgery persist should be left to determine for themselves the most ethical and effective ways of confronting the practice—without advice, but with support, from Western feminists.

Half a world away, Baylis, Jocelyn Downie, and Susan Sherwin chronicle the frustrations they encountered in trying to press their way into a Canadian public policy forum to ensure a gender-sensitive and inclusive research agenda. In the face of defeat, they reexamined their strategies and were compelled to acknowledge that well-crafted arguments are unlikely to persuade without supplementation by more explicitly political action. Despite disparities in the groups addressed—Jones's focus on grassroots movements and the Canadian group's concentration on the upper echelons of power—both situations reveal the need for strategies to resist treating women as though they were incapable of making responsible decisions about their own well-being. As Barbara Nicholas so persuasively points out in her sensitively wrought contribution, we need to hear many more stories of strategies that have worked—stories both by women near the centers of power and by those at the margins. Only through such dialogue can we hope to interject feminist concerns into the dominant discourses and practices that define the norms of medical care.

Notes

1. Laura M. Purdy, "Good Bioethics Must Be Feminist Bioethics," *Philosophical Perspectives on Bioethics*, ed. Wayne L. Sumner and Joseph Boyle (Toronto: Toronto University Press, 1996), 145.

2. Susan Sherwin, "Feminist and Medical Ethics: Two Different Approaches to Contextual Ethics," in *Feminist Perspectives in Medical Ethics*, ed. Helen Bequaert Holmes and Laura Purdy (Bloomington: Indiana University Press, 1992), 29, note 6.

3. Samuel Gorovitz, Andrew L. Jameton, Ruth Macklin, John M. O'Connor, Eugene V. Perrin, Beverly Page St. Clair, and Susan Sherwin, *Moral Problems in Medicine* (Englewood Cliffs, NJ: Prentice-Hall, 1976).

4. Boston Women's Health Book Collective, *Our Bodies, Ourselves* (New York: Simon & Schuster, 1969).

5. Holmes and Purdy, eds, *Feminist Perspectives*, editors' introductions.

6. Drucilla Cornell, *The Imaginary Domain* (New York: Routledge, 1995); and Moira Gatens, *Imaginary Bodies* (New York: Routledge, 1996).

Part I
Redirecting Bioethical Theory

1

Rehabilitating Care

Alisa L. Carse and Hilde Lindemann Nelson

The ethic of care validates skills and virtues traditionally associated with women and women's roles. This presents feminists in particular with a dilemma. On the one hand, there is a vital need for an ethic that takes the experiences of women seriously, and the ethic of care does just that, capturing certain features of our moral lives that other, more standard approaches to morality underplay or ignore. On the other hand, the ethic threatens to support and sustain the subordinate status of women in society, contributing to the exploitation and denigration of women with which feminist ethics is more broadly concerned. If it can be shown that the threat is real, feminists may feel obliged to repudiate the approach.

Can the ethic of care be saved? In this chapter we examine four standard complaints that are lodged against the care orientation: (1) complaints concerning the exploitation of caregivers; (2) the challenges of caretaking to caregiver integrity; (3) the potential perils of conceiving the mother–child dyad normatively as a paradigm for human relationships; and (4) the inadequacy of the care ethic in securing social justice among relative strangers. We find that much can be done to address all four complaints and that progress can thus be made toward rehabilitating care. We close by discussing why and how the ethic might be further developed.

Self-Sacrifice: The Problem of Exploitation

Broadly construed, the care ethic poses a challenge to prevailing models of moral knowledge and responsibility, especially the tendency in

17

18 A. Carse and H. Nelson

ethical theory to construe as paradigmatic those forms of judgment and response that abstract away from the concrete identities of others and our relationships to them. An adequate grasp of the moral contours of specific situations, especially as they concern other people and our responsibilities to them, requires an acute attentiveness to particularity and to the situation-specific nature of others' needs. As Margaret Walker writes, "The others I need to understand . . . are actual others in a particular case at hand, and not repeatable instances or replaceable occupants of a general status" (1989, 17). Such attention to particularity competes uneasily and sometimes irreconcilably with the movement toward abstraction and generalization characterized by the impartial, principled deliberation and justification emphasized in dominant moral theoretical models, particularly those rooted in Kant and liberal contractarianism.

The rejection of impartiality and principlism that characterizes much care-oriented ethical theory is one source of the criticisms that have been raised against care theory. The emphasis within care theory on attunement to concrete others, their situations, and their relationships to us is rooted in an axiology centering on our moral responsibility to understand and attend to the needs of others. Care theory's insistence on the importance of acute sensitivity to concrete circumstances, and on the cultivation of those capacities necessary for sympathetic, imaginative projection into the perspectives and situations of others, derives from the view that such sensitivity and the capacities it entails are crucial to achieving an understanding of the way others see and experience their needs, and thus to the ability to care for them effectively (Gilligan 1982; Held 1993; Noddings 1984).

This raises the first set of worries. Though care-oriented theory is permeated with metaphors of perception, the activity of caring can be obtuse and indiscriminate. A conception of the care ethic that includes no general normative constraints to regulate its force or direct it toward worthy objects only reinforces existing stereotypes of selfless, womanly sacrifice. If we are to protect against exploitative, abusive, demeaning, or otherwise unfair patterns of distribution and responsibility in our roles and relationships, we must have some way of reflecting critically on our roles and relationships to determine which of the expressed needs, expectations, and demands confronting us are morally legitimate. And since we cannot respond to every legitimate claim on our care, we require a means of distinguishing between the care we *may* give and the care we *must* give.

Some would argue that the ethic of care is an ethic of intimate relationships in which there is no need for justice. We will suggest later that this circumscription of the care ethic underestimates its potential.

The point we want to emphasize here is that care giving that is servile or exploitative should not be morally condoned in *any* relational context. The scope of justice properly extends not only to formal relations among distant strangers, but also to our more proximate relational spheres. Since family members and friends (or proximate strangers) differ in strength, power, and degrees of dependency, the vulnerability invited by intimacy and mutual concern can make these relationships particularly susceptible to problems of exploitation and injustice.[1] As Jean Hampton warns, "A moral agent has to have a good sense of her own moral claims if she is going to be . . . a real partner in a morally sound relationship. She must also have some sense of what it is to make a legitimate claim if she is to understand and respond to the legitimate claims of others and resist attempts to involve herself in relationships that will make her the mere servant of others' desires" (1993, 231).

To facilitate the task of recognizing a servile relationship—whether "in the family, the marketplace . . . political society, or the workplace"—Hampton herself looks to a justice perspective and proposes a "contractarian test." She invokes David Gauthier's claim that our sociality "becomes a source of exploitation if it induces persons to acquiesce in institutions and practices that but for their fellow-feeling would be costly to them" (Gauthier 1986, 11; quoted in Hampton 1993, 239). In assessing our relationships, we are to ask whether we could "reasonably accept the distribution of costs and benefits . . . that are not themselves side effects of any affective or duty-based ties between us . . . if [the relationship] were the subject of an informed, unforced agreement in which we think of ourselves as motivated solely by self-interest" (Hampton 1993, 240).[2] The point is not to ensure relations of equal reciprocity where this is impossible, but to prohibit relational arrangements in which one party exploits another by *taking advantage of* his or her affections (ibid. 247–248). This test is intended in effect to introduce the Kantian constraint that no member of a relationship be treated as a mere instrument of others.

Now, there is no question that an ethic emphasizing responsiveness to others' needs risks reinforcing our vulnerability to others' unfair demands. Indeed, it is precisely in the interest of setting just limits on our obligations to others that many critics of care theory insist that acceptable forms of caring presuppose that prior conditions of justice are met. We want, however, to suggest ways in which care theory itself can enrich our notions of relational justice, thereby better accommodating the different kinds of relationships and roles we inhabit.

It is clear that Hampton's contractarian test can at best offer limited guidance. As she herself notes, it does not apply in relationships of "radical inequality," such as those between a parent and infant, or a caretaker and someone seriously ill or infirm—relationships in which one party is "incapable of reciprocating the benefit[s]" provided by the other. And yet many of our most demanding relationships are of just this sort. Moreover, there is a wide range of relationships between those of "radical inequality," on the one hand, and material equality, on the other. Consider relationships of material inequality in which limited forms of reciprocity are possible, especially forms of affectionate expression and exchange. An adult who cares for her ill father may well derive a significant benefit from the relationship through the love and appreciation he expresses, though he may be too incapacitated by illness to return caretaking of the same kind. Both the quality of the affection and the balance of affection exchanged should play a role in our moral assessments of the justice of such relationships.

A further point: the caregiver–recipient dyad rarely exists in isolation from other relationships; the caregiver is generally nested in a cluster of relationships from which she in turn can draw care. Suppose an individual cares for her sister, whose Alzheimer's disease has progressed beyond the point where even affection is possible. If her family expects her to do the labor of care simply because she is a wife or a mother, or because they don't want to do it themselves, or because they are indifferent to her situation, she is arguably exploited, not because she gets nothing in exchange for the care she gives or because she yields to her sister's needs out of affection, but because capable members of her family ignore the burdens they impose on her by failing to be concerned with *her* welfare and thus to do their share. To argue along these lines is to see the fact of familial relationship to have at least prima facie import for the care we owe others. We will return to this point below.

An acceptable account of relational justice must require respect, concern, and support for the well-being and flourishing of *all* parties within relational structures. Respect from a care perspective will, as Robin Dillon argues, see "individuals as equally worthy of attention, consideration, and concern, equally worthy of understanding and care" (1992, 122). On this view, justice prevails when each is attended to, each is heard, each is recognized, and—crucially—no one's welfare is ignored or dismissed. The able husband who tenderly acknowledges how hard his wife works but never lends a hand to help fails to manifest the respectful concern this understanding of justice would demand, for he does not actively support his wife's flourishing.

This suggests that it is not sufficient, in ensuring relational justice, that one refrain from degrading others or from treating others as mere instruments to one's own ends; within a care orientation, respectful caring will require the mutual, active promotion of one another's well-being. As Dillon writes, such caring joins "individuals together in a community of mutual concern and mutual aid" (129).

This notion of respect differs from the Kantian notion: people are to be valued and respected directly, as concrete, particular selves, not because they are taken universally to possess an abstract and generic capacity for rational autonomous agency. Moral focus is placed on individuals' idiosyncracies and vulnerabilities and on the quality and particularity of specific interpersonal relationships; the fact of human interdependency is recognized as morally fundamental. Within a care ethic, a decent solution to the problem of exploitation will refer us both to an examination of the balance of relational goods exchanged by individuals (including affection, concern, humor, and the like) and to the broader structure of the relational networks in which we live. It will challenge us to reflect on the care we may give when our proximate relationships are marked by respectful love and to explore further how facts of relational distance and proximity ought bear on the care we must give as a matter of relational justice.

Self-Effacement and the Problem of Integrity

The need to set limits on care is an important problem for the care ethic not only because of the danger of exploitation, but also because of the danger of oppressing the recipient of care. The imposition of care on another without consulting her wishes or trying to understand her needs from her own point of view is rightly excoriated as paternalism; when we care for another solely on our own terms, we act arrogantly. The desire to avoid arrogant and invasive caring prompts many care theorists to insist on the requirement that the caregiver enter imaginatively and empathically into the world of the person being cared for.

As Nel Noddings writes, caring "involves stepping out of one's own personal frame of reference into the other's" (1984, 24). As caretaker, "I set aside my temptation to analyze and to plan. I do not project; I receive the other into myself, and I see and feel with the other" (30). The caretaker actively assimilates the other's values and ideas and "affectively interiorizes" an alien perspective, supplanting her own. She undergoes what Noddings calls "motivational displacement," in which she grasps what the person receiving her care wants for himself and allows that want to supersede her own motives for action (Noddings 1984, chap. 2); she also adopts the interpretive framework of the other, yielding her own understanding of the situation and its moral stakes.

But this motivational and interpretive "displacement" raises a second set of problems. As crucial as the skills of receptivity are to care-taking, they can pose serious risks in their own right—among other things, the risk of morally problematic self-effacement. Sandra Bartky (1990) offers a trenchant exploration of ways in which care giving can threaten caregivers with what she calls "epistemic" and "ethical dis-empowerment."[3] She writes:

> An "epistemic lean" in the direction of the object of her solicitude is part of the caregiver's job. . . . There is, then, a risk for women's epistemic development in our unreciprocated care giving. . . . Many of us, I suspect, have been morally silenced or morally compromised in small ways because we thought it more important to provide emotional support than to keep faith with our own principles. . . . More corrosive is a danger that inheres in the very nature of intimate care giving—the danger of an "ethical lean" that, like the epistemic lean, . . . may rob the caregiver herself of a place to stand. (1990, 111–113)

Bartky does not deny that caretaking has its attendant powers. Indeed, one may be and feel tremendously efficacious as a healer, consoler, and sustainer, the source of "a great reservoir of restorative power" (113). This, she claims, helps explain why women, among others, have persisted with little protest in roles and relationships that have exacted from them high epistemic and ethical costs. Her example is that of Teresa Stangl, who, though herself devoutly religious and opposed to Nazism, paid homage to her husband, Fritz, comforting and succoring him in ways that made it possible for him to continue in the capacity of commandant at Treblinka, even while she found his work morally repugnant.

Less dramatic but more familiar is the example of the health professional—the nurse, who, in wrestling with the conflict between her patient's desires and her own sense of what is right, can resolve the conflict with her conscience only by quieting it. If she takes herself to be obligated to step outside her own moral beliefs and commitments in order to serve her patient well, she may thereby be engaging in moral and ideational modes that reinforce her self-estrangement and subservience as a caretaker (Bartky 1990, 117).

Responsiveness to another's needs, however, ought not involve over-identification with the other, or assumed possession of the other's condition. Indeed, this can compromise care, by bringing one's focus onto oneself—one's own fears, felt needs, or aspirations. It thus risks being disrespectful of the other as a separate locus of experience and agency. A viable ethic of care must address the profoundly important moral challenge of healthy receptivity. It must acknowledge how traits that are in some contexts clearly virtues can be distorted and transformed, how tenderness, imagination, empathy, and sympathy—though often essential to caring effectively for others—can chip away at the integrity of the caregiver, promote moral compromise or even complicity in moral wrongdoing, and thus be potentially both destructive of the caregiver and morally pernicious in their effects.[4]

One theoretical solution to the problem is to incorporate self-care into the ethic of care and so keep it from being an ethic of self-efface-ment. The success of this solution, however, depends on the theoretical motivation for self-care. If the point of self-care is to enable the care-giver to care better for others, self-care is, as Rosemarie Tong observes, nothing more than "a disguised form of other-directed care" (1993, 128). If the one who cares has dedicated herself to another so com-pletely that she sets even care of herself at the other's service, she stands in danger of losing herself. For this reason, self-care must have its ethical basis in an affirmation of the caregiver's integrity.

Cheshire Calhoun (1995, 257) has recently suggested that integrity is not only a personal virtue, by which one acts to protect the boundaries of the self, but also centrally a social virtue: it involves standing for something *to other people*. As one among many deliberators, each of us uses her best moral judgment to answer the question "What is worth doing?" We try to answer this correctly not only for ourselves, but "for, and before, all deliberators who share the goal of determining what is worth doing."

By offering us a picture of integrity that reminds us that we are social selves, selves in relationship, Calhoun builds accountability to others directly into her understanding of what it is to "be oneself." Being one-self is just being responsible to, and before, others for one's moral judg-ments. And this requires that one take one's own judgments seriously (257). A self of the kind presupposed by Calhoun's picture of integrity cannot easily be lost by empathic caring for others, because it is the self of a morally developed person. When we hold the other's felt needs or desires to be deeply at odds with our own conception of the good, or of the other's interests, integrity requires that we follow our conscience and refuse to yield to the other's demands. However, it also requires that we take the other person's perspective seriously, consider it openly, and understand its potentially important connection to her self-conception, even if we reject it (Dillon 1992, esp. 125).

Receptivity to the other must not be confused with self-abnegating absorption into the other. An empathetic caretaker must be able to sur-vive as a strong, intact, self-respecting person in her own right. If Cal-houn's account of integrity is correct, the nurse who faces the problem of doing something morally questionable because her patient desires it must ask, "How, when I care for others, can I stand for what in my considered opinion is ethical care?" She thus represents both to her patient and to herself her own best judgment of what is worth doing.

Material Inequality: The Problem of the Mother–Child Paradigm

Some who write in care theory take its moral paradigm to be relationships of material inequality and unidirectional dependency, such as that found in the mother–child dyad. For Noddings (1984), for example, the ideal image of care—the mother nursing her child—is unidirectional care that asks for nothing in return. Sara Ruddick (1989) looks to a normative account of the structure of maternal practice for a characterization of virtues and modes of being that are crucial to fostering the preservation, growth, and socialization of a child. She sees a great need to cultivate in people such traits as "attentive love," "resilient cheerfulness," innovation, and responsiveness to others in their singularity. Virginia Held, too, focuses on the mother–child dyad in her examination of the moral contours of relationships characterized by nonvoluntary participation, mutual support, dependency, and irreplaceability. She argues that marketplace models of "rational man," on which ethical concerns pertain to constraints on voluntary, cooperative interaction and ways of ensuring mutual noninterference, cannot adequately address the morality of relationships of inequality and dependence (1987).[5]

Many critics have rejected maternal–child (and other familial) models of moral relationship, claiming that, at best, they represent perilous paradigms. Unrequited care, romanticized as a model for human relationships, can only reinforce oppressive roles and practices, teaching those who are cared for to receive without giving and doing nothing to promote the self-respect and equal moral standing of the caretaker (Nelson 1992, 10; Tong 1993, 128; Hoagland 1991, 254; Card 1990, 106). It is also argued that the forms of material inequality, unidirectional dependency, and asymmetrical concern that characterize relations such as that between a mother and her child are to be outgrown or transcended rather than celebrated.[6]

Moreover, it has been pointed out that construing unequal, emotionally engaged dependency relationships as morally paradigmatic fails to provide adequate moral protection against arbitrary and abusive uses of caretaker power; we need to be on guard against the potential for cruel or disabling exercises of maternal authority and discretionary leeway (Hoagland 1991, 252; Tong 1993, 125). It is precisely the asymmetries of power, authority, and vulnerability in parent–child relationships that lead many to deem them ill-suited as moral paradigms, for they permit (even require) forms of domination that are morally inappropriate in relations among equal adults (see, e.g., Grimshaw 1986, 215; Allen 1986; Code 1987; Houston 1987).

It would, of course, be a mistake to see the mother–child dyad as a sole and sufficient moral paradigm. But it does not follow that there is no moral wisdom to be gained from normative models of relationships (such as mother–child) that are in fact characterized, often ineluctably, by inequalities of power, knowledge, vulnerability, and dependency. We need not think of such relational models as exhaustive in order to take them seriously as crucial sources of moral insight and important objects of ethical attention. As Annette Baier has argued, the relative weight given to relations among "equals" in moral theory has led to a damaging "moral myopia." Baier writes that modern moral philosophers have managed

> to focus their philosophical attention . . . single-mindedly on cool, distanced relations between more or less free and equal adult strangers. . . . Philosophers who remember what it was like to be a dependent child or who know what it is like to be a parent or to have a dependent parent, an old or handicapped relative, friend, or neighbor will find it implausible . . . [to] see morality as essentially a matter of keeping to the minimal moral traffic rules, designed to restrict close encounters between autonomous persons to self-chosen ones. . . . For those most of whose daily dealings are with the less powerful or the more powerful, a moral code designed for those equal in power will be at best nonfunctional, at worst an offensive pretense of equality as a substitute for its actuality. (1994, 114–116)

The fact remains that we are, as humans, deeply interdependent; it is not uncommon to be in need of care and expertise that we cannot reciprocate fully or at all. The vulnerability we experience when we are ill, for example, is often intensified by our lack of medical expertise: we must hand our lives and welfare over to others who have knowledge, training, and experience we do not have. We are in a similar state of dependence when we submit to the judgment and skill of pharmacists, nurses, train conductors, or the teachers who oversee our children each day.

Although we can retain broad rights of refusal and request, contract models for relationships of material inequality are inherently impoverished, given the material realities of need, on the one hand, and limited knowledge and expertise, on the other. We are often not in a position to negotiate the terms of our relationships as equal partners; we must yield discretionary power to others. Nor can we rely on negotiation to assure that our welfare will be sought (Baier 1994, 116). Relationships with those who care for us or for our loved ones, those on whose wisdom, skill, and goodwill we depend, must of necessity be bound by *trust* (Pellegrino 1991). When we are in conditions of dependency, we must often place presumptive faith in those entrusted with our welfare, relying on them not to exploit our vulnerability or to abuse the power we have given them (Baier 1987; Baier 1994, chaps. 6, 7, 8, and 9).

This is not to deny that some of the material inequalities and imbalances that structure our relationships are themselves morally suspect and avoidable, and thus are to be overcome. Nor is it to maintain that we should simply trust all who come our way. It is, rather, to insist that our moral models and guides realistically acknowledge the full extent of our mutual interdependence as human beings and to suggest that a large measure of merited trust is necessary to the survival and flourishing of all of us, most of the time (Baier 1994).

This, in turn, raises crucial questions for political morality, concerning how our practices and institutions succeed or fail to sustain trustworthy relations among us.

The Scope of Care: The Problem of Strangers

Many critics argue that care, which is naturally extended to intimates and others with whom we come into direct contact, cannot help us to resist the evil that strangers do to strangers. Claudia Card notes, for example, that when all of morality is subsumed under the care we provide to our families, friends, and others in proximity, too much of the world is left out: we are too easily tempted to sexism, racism, xenophobia, homophobia, and disregard for future generations (1990, 101–108). In the absence of principles that can show us toward whom our care ought to be directed, we can care only for those with whom we happen to be in relation, and although a care ethic might bid us individually to care for those social causes we find most worthwhile, it cannot help us revise the institutions and ideological and economic forces that play a large role in such evils as world hunger or homelessness. These endemic ills require radical reconfigurations of social structures. A care ethic alone cannot address all of morality. Because it focuses on the particularistic and the personal, it is especially well designed to do the work of fine detail, but it is too delicate an instrument for some large-scale problems posed by social justice. Nonetheless, the ethic can play a role in addressing both the conditions of social justice and the directions we must take in reframing our conceptions of it.

In understanding how, it is useful to draw a distinction between "caring for" and "caring about" others.[7] "Caring about" can be seen to presuppose a position in value theory—What makes x worth caring about?—and this position need not be divided along intimate/stranger, private/public lines. I can care about the needs of future generations, the suffering of Bosnian Muslims, or the health of children far away and unknown to me personally. "Caring for," on the other hand, is an activity involving moral skill; we must decide how best to care for what we care about and then act accordingly. Marilyn Friedman (1987, 103) gives examples of what care might look like outside of more intimate spheres: "In its more noble manifestation, care in the public realm would show itself . . . in foreign aid, welfare programs, famine or disaster relief, or other social programs designed to relieve suffering and attend to human needs" (see also Tronto 1993, chaps. 1, 5, and 6).

Justice is undermined, among other things, by forces of bigotry and hatred, by competitive desire for unilateral advantage, by more seemingly innocuous forms of "tunnel vision" that emerge from indifference toward others or ignorance about them, and by the many often unconscious and unacknowledged aversions and biases that are in fact at play in our resistance to receiving one another in welcome partnership (Young 1990). These include prejudices about "funny" accents; discomfort with scarred faces, personal odors, or body types; and negative preconceptions about race, ethnicity, religion, sexual orientation, and the like.

The harm we do to strangers often is founded not only in bigotry and hatred, but also in forms of indifference that render us uninclined even to treat them justly. The extension of justice requires that others be on our "radar screen." They get there by being cared about.[8] An ethic of care that propels us to extend the scope of our attentiveness and responsiveness might go a long way (even if not all the way) toward undercutting crucial impediments to social justice. As Walker notes, we need to encourage just those virtues and capacities necessary "to defend ourselves against dispositions to keep strangers strange and outsiders outside" (1989, 23).

More fundamentally, while many have urged the importance of conceptualizing justice in ways that abstract from particulars of race, gender, religion, and the like in order to provide protection against prejudice and parochialism, the ethic of care challenges us to become attentive to differences in perspective and need *as a demand of justice*, to resist the human tendency to remain insensitive to and unconcerned about what is unfamiliar or relationally and personally distant. Crucial to this end is a normative conception of care that does not ground it solely in love and affection for those whose connections to us are visible, but that urges respectful, compassionate concern for the welfare of others even in the absence of bonds of affection or relational, geographic, or cultural familiarity.[9] We underestimate the care ethic when we confine it to dyadic relationships or to proximate spheres of social relationship and interaction.

Why Should We Care?

Much of our discussion has assumed that the ethic of care introduces important dimensions into ethical discourse. The question we need to pursue further is why we should regard the care ethic as a serious moral contender. It is not sufficient to claim that it reflects many women's lived experiences in ways that standing moral theories cannot, as this invites the objection that something might be wrong with women's lives and the norms and demands structuring them, rather than with standing theories and the norms and demands they set out. Such a claim need not blame women; it might simply express the view that women have been so oppressed and their lives so distorted that little of moral value can be gained from studying them. We must therefore ask whether there is an argument independent of a simple gender-bias thesis for maintaining that the care perspective and the norms and standards it engenders ought to have a claim on our attention, ought to be granted moral legitimacy.

To this question we reply that the ethic of care ought to be granted moral legitimacy because we need such an ethic if we are to be honest about what is required for human flourishing. Women's traditional labor—a vastly disproportionate share of the work of caring—has been important moral labor. It is labor that underscores the moral significance of human *interdependence*, raising ethical concerns about aloneness, abandonment, neglect, and isolation—concerns that arise especially when we are in special states of vulnerability, as are those who are young, ill, frail, disabled, or otherwise in need of others' care. Human thriving requires a world in which there are loving parents and caring citizens: trustworthy and responsible nurses, physicians, political representatives, teachers, neighbors, and cabbies. A stable social order depends on the presence in many of us, at least, of benevolence, compassion, kindness, imagination, and trustworthiness—traits that diminish conflict, promote cooperation, and secure the nurturance we all need. In the absence of a robust ethic of care, no one would ensure that children were tended and educated, that the needy and powerless were protected against neglect and abandonment, that we would receive attentive care when we were ill or downtrodden. Standard moral theories can continue to underplay these tasks of care only if they continue to ignore those whom Baier has dubbed "the long unnoticed moral proletariat . . . mostly female" (1987, 50), who do so much of the unacknowledged work on which our survival and flourishing depend.

Precisely because this work is so morally valuable, retaining a vivid sense of its potential dangers remains a crucial task for care theorists. We need to construct a frame of discourse that extends beyond the "lived experience" of the daily round of care, or we will find ourselves without the conceptual and theoretical resources required to protect women and other caregivers from further exploitation and disrespect.

Central to this reconstructive project will be the degendering of this ethic: the extension of its norms and prescriptions to men as well as women and to traditionally male domains of our social world. This will require, as Patrocinio Schweickart asserts, "the incorporation of the moral values traditionally associated with femininity into the authoritative discourses of the community, into the way we reflect on, judge, and articulate the worthiness or unworthiness of actions, relations, laws, and institutions" (1993, 188). It will engage us in critical reflection about many social and cultural practices and expectations concerning *who* takes on the responsibilities of caring for those in need, *under what conditions*, and *at what cost*. Given our social roles, practices, and institutions, how does the work of caretaking, which is necessary to society, affect the distribution of benefits and burdens within society? How, that is, does this work affect economic status and stability, job security and promotion, the ability to pursue projects of one's own, or to contribute as an equal in the constitution and evaluation of our shared moral life? It is only after we bring the moral work of caring to the table that we can begin to articulate unromanticized, realistic, and just conceptions of what this ethic should entail.

Notes

An earlier version of this essay first appeared in *Kennedy Institute of Ethics Journal* 6 (1996): 19–35. We are grateful to Margaret Walker, Betsy Poslow, and Maggie Little for helpful comments on earlier drafts of this chapter.

1. For rich discussions of justice in families, see Friedman (1987, esp. 102–103) and Ruddick (1995).

2. Presumably, "self-interest" is here to be defined in terms that make no reference to the interests of others, so as to avoid legitimizing servile desires or preferences an individual may take herself to have in employing such a test. See Gauthier on nontuistic preferences (1986, 311).

3. Bartky is looking at these issues through the lens of gender and gendered

divisions of labor, but her analysis yields a general insight for the potential perils of an ethic of care.

4. See Annette Baier (1994, esp. chaps. 6 and 7).

5. See Treblicot (1983) for an excellent collection of essays on mothering and feminism. Held (1987) and Nelson (1995) offer two different expositions of the view that chosen relationships are not paradigmatic of human connection.

6. Sarah Lucia Hoagland has argued, for example, that the mother–child dyad, far from being a moral paradigm, represents a morally *diminished* relationship, precisely because it is structured by unidirectional dependency and unequal reciprocity. Achieving independence is, she argues, a condition of attaining full moral development and maturity. It should therefore be a moral objective of caretaking relationships (e.g., parenting, teaching) "to wean the cared-for of dependency" when possible (Hoagland 1991, 250–251; Tong 1993, 124). We should not model our ethical relationships on dependency relationships that are "ideally transitory" (Hoagland 1990, 110; quoted in Tong 1993, 127).

7. This distinction was suggested by Virginia Sharpe in conversation. It has been made by Tronto (1993) as well.

8. Maggie Little invokes the example of whites who, while watching the news, do not notice the murder of a black teenager but look up in alarm when a white is killed.

9. For an insightful account along these lines, see Dillon (1992).

References

Allen, Jeffner. 1986. *Lesbian Philosophy: Explorations*. Palo Alto, CA: Institute of Lesbian Studies.

Baier, Annette. 1987. The need for more than justice. In Marcia Hanen and Kai Nielsen, eds., *Science, Morality and Feminist Theory*, supp. vol. 13 of *Canadian Journal of Philosophy*, (1987): 41–56.

———. 1994. *Moral Prejudices: Essays on Ethics*. Cambridge, MA: Harvard University Press.

Bartky, Sandra. 1990. *Femininity and Domination: Studies in the Phenomenology of Oppression*. New York: Routledge.

Calhoun, Cheshire. 1995. Standing for something. *Journal of Philosophy* 42 (5): 235–260.

Card, Claudia. 1990. Caring and evil. *Hypatia* 5 (1): 101–108.

Carse, Alisa. 1991. The "voice of care": Implications for bioethical education. *Journal of Medicine and Philosophy* 16: 5–28.

Code, Lorraine. 1987. Second persons. In Marcia Hanen and Kai Nielsen, eds., *Science, Morality and Feminist Theory*, supp. vol. 13 of *Canadian Journal of Philosophy* (1987) 357–382.

Dillon, Robin S. 1992. Care and respect. In Eve Browning Cole and Susan Coultrap McQuin, eds., *Explorations in Feminist Ethics: Theory and Practice*. Bloomington: Indiana University Press.

Friedman, Marilyn. 1987. Beyond caring: The de-moralization of gender. In Marcia Hanen and Kai Nielsen, eds. *Science, Morality, and Feminist Theory,* supp. vol. 13 of *Canadian Journal of Philosophy* (1987) 87–110.

Gauthier, David. 1986. *Morals by Agreement.* Oxford: Oxford University Press.

Gilligan, Carol. 1982. *In a Different Voice: Psychological Theory and Women's Development.* Cambridge, MA: Harvard University Press.

Grimshaw, Jean. 1986. *Philosophy and Feminist Thinking.* Minneapolis: University of Minnesota Press.

Hampton, Jean. 1993. Feminist contractarianism. In L. Antony and C. Witt, eds., *A Mind of One's Own.* Boulder, CO: Westview Press.

Held, Virginia. 1987. Non-contractual society. In Marcia Hanen and Kai Nielsen, eds., *Science, Morality, and Feminist Theory.* supp. vol. 13 of *Canadian Journal of Philosophy* (1987) 111–137.

———. 1993. *Feminist Morality: Transforming Culture, Society, and Politics.* Chicago: University of Chicago Press.

Hoagland, Sarah Lucia. 1991. Some thoughts about "caring." In Claudia Card, ed., *Feminist Ethics.* Lawrence: University Press of Kansas, 246–263.

Houston, Barbara. 1987. Rescuing womanly virtues: Some dangers of moral reclamation. In Marcia Hanen and Kai Nielsen, eds., *Science, Morality, and Feminist Theory,* supp. vol. 13 of *Canadian Journal of Philosophy* (1987) 237–62.

Nelson, Hilde Lindemann. 1992. Against caring. *Journal of Clinical Ethics* 3 (March): 8–15.

———. 1995. Dethroning choice: Analogy, personhood, and the new reproductive technologies. *Journal of Law, Medicine & Ethics* 23 (2): 129–135.

Noddings, Nel. 1984. *Caring: A Feminine Approach to Ethics and Moral Education.* Berkeley and Los Angeles: University of California Press.

Pellegrino, Edmund D. 1991. Trust and distrust in professional ethics. In Edmund D. Pellegrino, Robert M. Veatch, and J. P. Langan, eds., *Ethics, Trust, and the Professions: Philosophical and Cultural Aspects.* Washington, DC: Georgetown University Press, 69–89.

Ruddick, Sara. 1989. *Maternal Thinking.* Boston: Beacon Press.

———. 1995. Injustice in families: Assault and domination. In Virginia Held, ed., *Justice and Care: Essential Readings in Feminist Ethics.* Boulder, CO: Westview Press: 203–223.

Schweickart, Patrocinio. 1993. In defense of femininity: Commentary on Sandra Bartky's *Femininity and Domination, Hypatia* 8 (1): 178–191.

Tong, Rosemarie. 1993. *Feminine and Feminist Ethics.* Belmont, CA: Wadsworth Publishing.

Treblicot, Joyce, ed. 1983. *Mothering: Essays in Feminist Theory.* Totowa, NJ: Rowman & Allanheld.

Tronto, Joan. 1993. *Moral Boundaries: A Political Argument for an Ethic of Care.* New York: Routledge.

Walker, Margaret. 1989. Moral understandings: Alternative "epistemology" for a feminist ethics. *Hypatia* 4 (2): 16–28.

Young, Iris Marion. 1990. *Justice and the Politics of Difference.* Princeton, NJ: Princeton University Press.

2

Just Caring About Maternal–Fetal Relations: The Case Of Cocaine-Using Pregnant Women

Rosemarie Tong

What Makes an Ethics of Care Feminist?

Alisa Carse, among others, has clearly captured the essence of an ethics of care. In its commitment to what Carse terms "qualified particularism," an ethics of care is, first, "an ethical orientation highlighting concrete and nuanced perception and understanding—including an attunement to the reality of other people and to the actual relational contexts we find ourselves in."[1] Second, in its commitment to what Carse terms "affiliative virtue," an ethics of care "asserts the importance of an active concern for the good of others and of community with them, of a capacity for sympathetic and imaginative projection into the position of others, and of situation-attuned responses to others' needs."[2] In other words, unlike an ethics of justice, an ethics of care resists the drive toward abstraction and universalization that most impartial, principled deliberation seems to require. An ethics of care is directly concerned neither with doing duty for duty's sake nor with maximizing the good of the aggregate; rather, it is focused on attending to the specific needs of particular individuals and on weaving thick webs of human relationships and responsibilities.

As I have just described it, this ethics of care is not yet a *feminist* ethics of care. Like Carse, I agree that most advocates of a feminist ethics of care believe that Western ethics is lopsided. Because men have been valued as a gender, everything associated with men—"independence, autonomy, intellect, will, wariness, hierarchy, domination, culture, transcendence, product, asceticism, war and death"[3]—has also been valued. In contrast, because women have been disvalued as a gender, everything associated with women—"dependence, community, connection, sharing, emotion, body, trust, absence of hierarchy, nature, immanence, process, joy, peace, and life"[4]—has also been disvalued. As a result of this sort of gender bias, Western ethics emphasizes the "male value" of justice as a morally better value to cultivate than the "female value" of care, an emphasis that feminist advocates of care aim to adjust by demonstrating that care is no less a moral virtue than is justice.

Although I think it is absolutely essential that culturally linked feminine values be affirmed just as loudly as culturally linked masculine values, like Carse, I do not think that what ultimately makes an ethics of care feminist is simply its celebration of culturally feminine moral values. Rather, I concur with Alison Jaggar that what ultimately makes any ethics, including an ethics of care, feminist is that it (1) proceeds on the assumption that women and men do not share precisely the same situation in life; (2) offers action guides "that . . . tend to subvert rather than reinforce the present systematic subordination of women";[5] (3) provides strategies for dealing with issues that arise in private or domestic life; and (4) "take[s] the moral experience of all women seriously, though not, of course, uncritically."[6] In short, what makes an ethics of care feminist is not only that it celebrates values like caring, but also that it refuses to permit a value like caring to "trap" women by requiring them, but not men, to tend others.

Advocates of a feminist ethics of care insist that unless "ethics" acknowledges that women's traditional virtues are just as necessary for full moral development as are men's traditional virtues, most men will not embrace the values of an ethics of care, even if most women embrace the values of an ethics of justice. As a result, women will continue to be the primary caregivers of the young, the old, the infirm, the disabled, and the distressed, and men (together with some beleaguered, "double-day" women) will continue to be the major "movers and shakers" in the realms of business, law, medicine, education, and government. Therefore, any ethics of care that can safely and accurately be termed "feminist" must distribute the weight of moral responsibility equally between the genders, not only by requiring women to be just as fair-minded as virtuous men supposedly are, but also by requiring men to be just as attentive to other people's needs as virtuous women supposedly are. What's more, both men and women must question whether the values that have been culturally associated with their respective genders are *true* moral values. Why, after all, should any human being embrace a "value" such as domination—a "value" that certainly does not *seem* to honor human beings' fundamental interests in freedom and well-being? And why, after all, should any human being enthusiastically affirm connection as a "value," when connection is sometimes used to attach not merely thin strings, but even heavy ropes, to people?

To be sure, if the aim of a feminist ethics of care is to achieve this type of gender equity, it will need to range over the public as well as the private realm. Advocates of a feminist ethics of care must regularly sit at the policy table as well as the kitchen table, for it is at the policy table that they will have an opportunity to shape the systems and structures within which men and women live and relate. In order to demonstrate that a feminist ethics of care not only *should* but also *can* move into the public realm, I propose to focus on one matter of public concern—namely, how to deal with cocaine-using pregnant women who, on account of their "habit," threaten harm to the fetuses they intend to bring to term. What I shall argue is that advocates of a feminist ethics of care will resolve this issue differently from either advocates of an ethics of justice or advocates of a nonfeminist ethics of care.

The Justice of Punishing Cocaine-Using Pregnant Women

There are at least four reasons why advocates of an ethics of justice target cocaine-using pregnant women, as opposed to alcohol-using or tobacco-using pregnant women, for punishment. First, unlike alcohol and tobacco, cocaine is an illicit and generally socially unacceptable drug. While jealously safeguarding its right to use alcohol and tobacco, mainstream society condemns cocaine users' habit as dirty, disgusting, destructive, and degenerate. Second, cocaine-exposed infants suffer in a particularly visible manner. For several months after birth, they typically shake uncontrollably (presumably an effect of some prebirth nerve damage); they also screech incessantly, apparently in painful reaction to even the gentlest of human touches. Some are severely brain damaged; many are subject to behavior disorders, coordination problems, and learning disabilities; and almost all are more functionally limited than they would have been had their mothers not been cocaine users.[7] Third, the number of cocaine-exposed infants is on a particularly dramatic increase. About one out of every ten U.S. newborns— 375,000 a year—is reportedly exposed in utero to one or more illicit drugs, most often cocaine. Many public urban hospitals claim that 20 percent of the infants they deliver show the effects of illicit drugs.[8] Fourth, the social and economic costs of dealing with cocaine-exposed infants and children are very high. Cocaine-exposed infants require longer than normal hospital stays and more expensive medical treatment than normal newborns; as they mature, they require more remedial educational programs and alternative social systems (preeminently, foster care) than normal children.[9]

Concerned about the plight of cocaine-exposed infants, advocates of an ethics of justice reason that if society condemns as child abusers parents who inject harmful substances into their children's bodies, it should also condemn as child abusers women who "deliver" cocaine to their fetuses in utero. Employing arguments that balance the rights and well-being of cocaine-exposed fetuses against the rights and well-being of cocaine-using pregnant women, advocates of an ethics of justice maintain that considerations about women's freedom must yield to considerations about fetuses' well-being.[10] Pregnant women who knowingly risk substantial harm to the fetuses they intend to bring to term are, in the estimation of those who seek to punish them, as worthy of criminal punishment as mothers who knowingly harm the children they have already birthed.

To be sure, most advocates of an ethics of justice concede that it would be a mistake for society to move from condemning as child abusers cocaine-using pregnant women (and possibly also alcohol-using pregnant women, who probably harm their fetuses even more than cocaine-using pregnant women harm theirs)[11] to condemning as child abusers pregnant women who smoke, fail to eat enough nutritious food, work in toxic environments, ingest too much caffeine, exercise too much or too little, or engage in too much sexual activity.[12] Not only are the consequences of the latter behaviors less clearly harmful than those of using cocaine, it is also inadvisable for the state to make pregnancy so onerous that women simply refuse to get pregnant or routinely abort their fetuses for fear of delivering them in anything other than a perfect condition. Nevertheless, advocates of an ethics of justice maintain that, within reasonable limits, the state *should* make women take their pregnancies seriously for the sake of themselves, their fetuses, and society as a whole. Everyone has a vested interest in seeing that children are born as healthy as possible.

The Injustice of Punishing Cocaine-Using Pregnant Women

In all fairness, the above characterization of the ethics of justice is misleading. For example, deontologists (most of whom espouse a retributive theory of punishment) base their case for punishing a person on that person's responsibility and blameworthiness for his or her actions. Thus, deontologists might not turn deaf ears to the point many feminists make: namely, that it is unfair for society to punish cocaine-using pregnant women for fetal abuse if it fails to provide these women with opportunities to overcome their addiction. Currently, rehabilitative centers for substance abusers are in short supply, and most of them are oriented to serving adult males.[13] They are ill-equipped to arrange day care for the older children of a cocaine-using pregnant women or to help her remove herself from the physically, psychologically, or sexually abusive situations that may be contributing to her use of drugs. Those centers that operate on an adult female–centered model and willingly accept clients tend to have extremely long waiting lists.[14] For all these reasons, deontologists might well agree with feminists that it is not appropriate to punish cocaine-using pregnant women. If a person is given no opportunity to overcome her addiction, and if her addiction is the cause of her acting in an irresponsible manner, then it is not fair to blame and subsequently punish her for her actions.

Deontologists should also be sympathetic to another point many feminists have made: namely, that as fetal-abuse sanctions have been applied so far, they have been applied in an uneven manner. In the first place, although men are sometimes just as guilty of fetal abuse as are women, pregnant women have been singled out as the targets of fetal-abuse legislation.[15] Why, ask feminists, it is just to punish a cocaine-using pregnant woman for fetal abuse but not also the man who insists that she use cocaine with him or else ("else" being a beating administered to the most vulnerable section of her anatomy)? Second, although there are many white cocaine-using pregnant women, it is almost exclusively African American cocaine-using pregnant women on welfare who have been brought up on a criminal charges for fetal abuse.[16] Many feminists suspect that because relatively affluent and still predominately white professionals—physicians, prosecutors, police officers—have little sympathy for those whose "lifestyles" depart most conspicuously from their own, they are more apt to view poor, minority women, as opposed to well-to-do white women, as irresponsible and selfish "junkies" who couldn't care less about the life within them.

To the degree that deontologists agree that, as currently applied, fetal-abuse sanctions are applied not only unevenly but also to women who either have no access to *good* drug rehabilitation programs[17] or are unable, *because* of their very addiction, to comply with their treatment regimes, they are not likely to support such punishments. Only if fetal-abuse sanctions are applied in a nondiscriminatory fashion to women who have it within their power to enter and abide by the terms of a good drug rehabilitation program will deontologists confidently apply the label "fetal abuser" to a cocaine-using pregnant woman who refuses treatment for her addiction. Whatever concessions a deontologist might make, he or she will tenaciously cling to the view that people who are not really to blame for their actions are not deserving of punishment.

Like deontologists, utilitarians (most of whom support the deterrence model of punishment) should be amenable to certain feminist arguments, including those of Dr. Helene Cole, who has argued forcefully that the threat of criminal punishment is unlikely to cause many cocaine-using pregnant women to overcome their habit. She notes that although cocaine-using pregnant women realize they can be sent to prison for using cocaine alone, most persist in using it. Given this fact, Cole concludes that whatever causes and reasons prompt cocaine-using pregnant women to ignore existing criminal penalties for cocaine use will prompt them to disregard new criminal penalties for fetal abuse. No matter how much concern they express for their fetuses' and children's well-being, and no matter how many opportunities they have to enter treatment programs, few will be able to break the hold cocaine has on their minds and bodies.[18] To be sure, some of these cocaine-using pregnant women *might* be able to overcome their addiction were the death penalty routinely imposed as the punishment for fetal abuse; but the imposition of such a draconian punishment on a particularly vulnerable group of women would probably not maximize society's long-term good even if it had its intended effect of deterrence.

Utilitarians should also listen to another point feminists make, namely, that punishing a cocaine-using pregnant woman who delivers a cocaine-exposed infant by imprisoning her and separating her—perhaps permanently—from her baby is unlikely to maximize the aggregate good. Not only has damage already been done to the child, but if the mother is not provided with good treatment for her addiction in prison, she will very likely return to using cocaine upon her release. If her child is then returned to her, she will have difficulty adequately parenting it; yet if her child is not returned to her, it will be shuttled from one foster home to another—a system of "parenting" that will not necessarily be an improvement on the care she might have provided.[19] Thus, by punishing cocaine-using pregnant women, society gains only an opportunity to express its fear and hate for women who, at least on the surface, look and act like *bad* mothers—that is, women who are not prepared to make any and all sacrifices for the welfare of their children.

Just and Caring Treatment Programs for Cocaine-Using Pregnant Women

Advocates of an ethics of justice who are amenable to certain feminist arguments about the unfairness and/or inutility of punishing cocaine-using pregnant women should be persuaded to endorse treatment programs for these women. Deontologists would be especially concerned that these programs foster the cocaine-using pregnant woman's autonomy and sense of personal responsibility for her decisions and actions. Utilitarians would, in contrast, be especially concerned that these programs not only help cocaine-using women overcome their addiction permanently, but also help them give birth to, and subsequently adequately parent, healthy children. Thus, one could easily imagine advocates of an ethics of justice supporting Grady Memorial Hospital's Project Prevent, an ambitious program funded by a $450,000 federal grant that cooperates with Atlanta's police department and homeless shelters to recruit pregnant drug users. Each woman receives personal attention from project adviser-advocates. The program also pays for transportation and other child care costs until the birth.[20]

Advocates of an ethics of justice might also find especially appealing the General Maternal and Child Health Project in Washington, D.C. Funded by a modest $110,000 grant from the Washington Junior League, as well as by some much smaller private donations, the hospital conducts a ten-week support program for patients drawn from its prenatal and drug-abuse wards. The women, many of whom have no homes, gather weekly for lectures. Each receives a nourishing meal and two gifts—one for herself and another for her infant—such as a blanket, baby clothes, or a car seat. Volunteer "godparents" urge their charges to stop taking drugs and to keep their doctors' appointments. So successful is this program that not one of the women who has participated in it has abandoned her child, a sign that she is assuming personal responsibility for the *healthy* child others have helped her deliver.[21]

Not only would advocates of an ethics of justice probably support the programs described above, so, too, would advocates of a nonfeminist ethics of care support them. To be sure, advocates of a nonfeminist ethics of care would probably insist that these programs be as individualized as possible and that genuine relationships be formed between the caregivers and the care-receivers. A genuine relationship might approximate the kind of human relationship Nel Noddings sketches in her book *Caring*. She defines a relation as "a set of ordered pairs generated by some rule that describes the affect—or subjective experience—of the members."[22] There are two parties in any genuine relation: the "one-caring" and the "cared-for." The one-caring is empathetically engrossed or "displaced" in the cared-for. The carer attends to the cared-for in deeds as well as thoughts, thereby making concern for the cared-for visible. In return, the cared-for actively recognizes the caring thoughts and deeds of the one-caring and responds to them by "turn[ing] freely towards his own projects, pursu[ing] them vigorously, and show[ing] his accounts of them spontaneously.[23] So, for example, if a volunteer godparent "keeps at" her charge, the relation is genuine if the cocaine-using pregnant woman *receives* this "nagging act" as a manifestation of care and says to herself something like, "It's time for me to start making my doctor's appointments. If my 'godparent' cares enough about me to do *x*, *y*, and *z* for me, the least I can do is show that I care for my baby."

Alternatively, a genuine relationship might approximate the mother–child relationship that Sara Ruddick claims is established through what she terms "maternal practice," a human activity that has three goals: preserving the life of one's child, fostering the child's growth, and making the child socially acceptable. In Ruddick's estimation, the kind of virtues necessary for maternal practice and the kind of thinking generated by it are precisely the ones caregivers should exhibit. In other words, the *ideal* caregiver helps the care-receiver develop into a person capable of giving as well as receiving care.[24]

Feminist Program for Cocaine-Using Pregnant Women: Just Caring

Although advocates of a feminist ethics of care would certainly prefer treating cocaine-using pregnant women in rehabilitative programs to punishing cocaine-using pregnant women in prisons, they would nonetheless have reservations about these programs. First, they would worry that the caregivers, most of whom are women, might be all too willing to go on caring even in the face of the purported care-receivers' nonresponsiveness to their efforts—to blame themselves for not trying hard enough in the event that their program turned out to be a colossal failure. Feminist advocates of an ethics of care would worry, in other words, lest female caregivers fall prey to the notion that caring is their duty whether or not anyone gives them anything in return. Thus, advocates of a feminist ethics of care insist that caregivers ask themselves the kinds of questions Sheila Mullett asks about caring: namely, is this the kind of caring that:

1. Fulfills the one caring
2. Calls upon the unique and particular individuality of the one caring
3. Is not produced by a person in a role because of gender, with one gender engaging in nurturing behavior and the other engaging in instrumental behavior

4. Is reciprocated with caring, and not merely with the satisfaction of seeing the ones cared for flourishing and pursuing other projects
5. Takes place within the framework of consciousness-raising practice and conversation[25]

Second, advocates of a feminist ethics of care would worry that the kind of care the care-receivers get is not *really* empowering them. As many feminists see it, the kind of care cocaine-using pregnant women need does not simply get them to see what is wrong with themselves and to mend the errors of their ways, or the kind that motivates them to please their caregivers by taking an interest in caring for their babies. Rather, it is the kind of care that also lets them see what social, economic, and cultural forces triggered their behavior in the first place and how they can join with other cocaine-using pregnant women to ask for the kind of assistance *they* think will really help them and their children have a better life. In other words, those of Iris Marion Young to be specific, a *feminist* ethic of care calls not for confession, but for the kind of consciousness-raising[26] that promises to deliver not only cocaine-using pregnant women, but all women, to that state of being Sara Lucia Hoagland has termed "autokoenomy." According to Hoagland, the autokoenomous woman is both separate from and connected to her community, the kind of person who depends on other people without being dependent on them.

> If I depend on someone and she is unable to meet her commitment, I may be set back, but I can carry on. Carrying on may involve replanning the scope of my venture, finding someone else to help me in the particular capacity my friend was unable to, joining with some larger project, or at times abandoning the project and moving on to something else. But I am the subject, still, of my choices. If, on the other hand, I am dependent on someone and she cannot keep her commitment, I fall apart: I cannot carry on; I am the object of events, the one to whom things happen. I am not the one who makes choices, and so I am not a moral agent.
>
> Further, if I am dependent on someone rather than depending on her, when over time things change and in evaluating her situation, she must also be aware that should she be unable to continue as before, I might fall apart. Thus she becomes responsible for me; and if she needs to slow down, she may respond with guilt feelings around my dependency when she really needs to focus her energy on herself for awhile.[27]

Any feminist who advocates an ethics of care *and* takes Hoagland seriously must resist the temptation to make those for whom she cares somehow dependent on her. Instead, she must have the courage to help herself as well as those in her care—in this instance, pregnant co caine-using women—to understand why addiction is more a social disease than a personal weakness. Only then will she be able to work together with her charges to oppose the structures and systems that contribute to women's disempowerment. Neither punishing nor treating cocaine-using pregnant women in a "maternalistic" way will empower them. What will empower them is the opportunity to challenge and change society's system. Advocates of a feminist ethics of care know that justice is not enough; however, they also know that caring is not enough. What is enough, in their estimation, is a combination of caring and justice that leads to equality on the macro level and autonomy on the micro level. In other words, the ultimate goal of a feminist ethics of care must be to create the conditions that empower all women—including cocaine-using pregnant women—to secure justice for one another and to care for one another and those to whom they wish to be related in ways that strengthen rather than weaken them.

Notes

1. Alisa L. Carse, "Qualified Particularism and Affiliative Virtue: Emphasis of a Recent Turn in Ethics," *Revista Medica de Chile* (1995): 10.
2. Ibid.
3. Ibid., 364.
4. Ibid.
5. Alison M. Jaggar, "Feminist Ethics," in Lawrence Becker with Charlotte Becker, eds., *Encyclopedia of Ethics* (New York: Garland, 1992), 366.
6. Ibid., 367.
7. Helene M. Cole, "Legal Interventions During Pregnancy," *Journal of the American Medical Association* 264, no. 2 (November 28, 1990): 2666–2667.
8. Anastasia Toufexis, "Innocent Victims," *Time*, May 13, 1991, 58–59.
9. Ibid.
10. Dawn E. Johnson, "The Creation of Fetal Rights: Conflicts with Women's Constitutional Rights to Liberty, Privacy, and Equal Protection," *Yale Law Journal* 95 (1986), 614–620.
11. Robert H. Blank, *Mother and Fetus: Changing Notions of Maternal Responsibility* (New York: Greenwood Press, 1992), 95–99.
12. Janet Gallagher, "Fetus as Patient," in Sherrill Cohen and Nadine Taub, eds., *Reproductive Laws in the 1990s* (Clifton, NJ: Humana Press, 1989), 201–202.
13. "Drug Abuse in the United States: A Policy Report," *Proceedings of the House of Delegates*, 137th Annual Meeting of the Board of the American Medical Association, June 26–30, 1988.
14. Cole, "Legal Interventions During Pregnancy," 2668.
15. Jean Reith Schroedel and Paul Beretz, "A Gender Analysis of Policy

Foundation: The Case of Fetal Abuse," in Patricia Boling, ed., *Expecting Trouble* (Boulder, CO: Westview Press, 1995), 91–95.

16. Henry Eichel, "S.C. Lawsuit Spotlights Debate on How to Treat Pregnant Cocaine Users," *Charlotte Observer*, January 31, 1994, 5A.

17. Cole, "Legal Interventions During Pregnancy," 2667.

18. Ibid.

19. Ken Garfield, "Babies Addicted: Moms Face Jail," *Charlotte Observer*, November 11, 1990, 1A.

20. Ann Blackman, "Mother-and-Child Reunion," *Time*, January 24, 1994, 58–59.

21. Ibid.

22. Nel Noddings, *Caring: A Feminine Approach to Ethics and Moral Education* (Berkeley: University of California Press, 1984), 3–4.

23. Ibid., 23.

24. Sara Ruddick, *Maternal Thinking: Toward a Politics of Peace* (Boston: Beacon Press, 1989), p. 17.

25. Sheila Mullett, "Shifting Perspectives: A New Approach to Ethics," in Lorraine Code, Sheila Mullett, and Christine Overall, eds., *Feminist Perspectives: Philosophical Essays on Method and Morals* (Toronto: University of Toronto Press, 1989), 119–120.

26. Iris Marion Young, "Punishment, Treatment, Empowerment: Three Approaches to Policy for Pregnant Addicts," in Boling, ed., *Expecting Trouble*, 123.

27. Sarah Lucia Hoagland, *Lesbian Ethics* (Palo Alto, CA: Institute of Lesbian Studies, 1988), 146.

3

Closing the Gaps: An Imperative for Feminist Bioethics

Helen Bequaert Holmes

I n what we call "health" care in the United States a widening gap separates the haves from the have-nots. The rich can have their lives prolonged by the most sophisticated medical technology in the world, while the poor often can't even get the minimum food and shelter that underpin health (Sherwin 1996, 56). Since the mid-1970s, the expansion of "managed care" has exacerbated this problem. The market emphasis of managed care forces physicians to spend less time per patient and to limit referrals and diagnostic aids (Miles 1997; Williams 1997). For-profit hospitals and health maintenance organizations (HMOs), the major form of managed care, are sprouting all over the United States and lining the pockets of entrepreneurs at the expense of the powerless (Evans 1997). More and more of the poor wait ever-longer hours to get into (or expelled from) hospital emergency rooms; more and more hospital stays are being cut shorter.[1] "The prevailing system [is] not simply . . . 'benignly neglectful' of women, minorities, and the poor, but . . . positively hostile to them" (Nelson and Nelson 1996, 355).

Yet another gap has been steadily widening: salaries of the chief executive officers (CEOs) of HMOs and insurance companies have become astronomical (Freudenheim 1995; Mathews 1995), while lower-echelon health care workers get laid off or are given heavier workloads and minuscule salary increases. In 1994, 34 percent of the costs in for-profit hospitals went to administration (Woolhandler and Himmelstein 1997).[2]

There is an insidious side to this gap. Our society reveres doctors;

45

we consider their advice more enlightened than anyone else's. But now managed care administrators "manage" doctors: physicians must toe the line, the *bottom* line. Our society may not realize what is happening, because administrators can bask in the reflected glow of our esteem for, and veneration of, doctors and in the cognitive and social authority of medicine.

In this chapter, I take a hard look at the relationship of these gaps to the field of bioethics[3] and at the external forces that have shaped that field. I argue that these forces, plus many of the same factors that led bioethics to pay so little attention to feminist work, have also induced that field to essentially ignore these widening gaps. I look at inconsistencies in bioethics' professed values and its failure to investigate moral problems connected with equality (Purdy 1996, 3), note patterns of domination and oppression (Sherwin 1996, 48), and consider "power relations of the persons who are involved in or who are affected by" the gaps (ibid, 52). A feminist approach does not merely analyze issues. Because it "strives to change the distribution and use of power" (Wolf 1996b, 8) and attempts to find "ways of reducing and, ultimately, eliminating oppression" (Sherwin 1996, 62), I shall offer some suggestions for reducing the gaps.

The Rise of an Elitist (but Powerless?) Bioethics

The twenty-year period of increasing health care disenfranchisement coincides exactly with the growth boom of the profession of bioethics. In the 1970s, bioethics was a plaything that a few academics pioneered as applied ethics when they weren't doing the more reputable theoretical ethics. Their salaries were as low as those of other academic philosophers.

But now look at bioethics! In the five-year period from 1982 to 1986 the percentage of large hospitals (i.e., 200+ beds) with ethics committees rose from 1 to 60 (Cohen 1988). By 1991, most state legislatures had mandated or recognized ethics committees (Wolf 1991, 807), and the number of bioethics think tanks had increased exponentially. By 1997, nearly every prestigious medical school had an ethics center or department. Faculty members there command salaries almost as high as their M.D. colleagues'. Indeed, the profession has now become so respectable that some physicians quit pediatrics and even surgery and cardiology to become full-time ethicists.

Bioethics has become an essential part of Western medicine even though it doesn't improve health. Ethics committees, ethics consultants, and now even attending physicians speak "ethics-ese." Bioethics' rising status as a profession is shown by its professional associations with high membership dues and by high registration fees for its conferences at posh hotels or resorts. You might think that moral concern about patient well-being would qualify one for participation, but it doesn't. Some meetings are by invitation only. Indeed, the profession's strong gatekeeping activities themselves raise ethical questions.

Who actually gets onto an ethics committee? Stuart Youngner and colleagues (1983) found that a committee of eight (the median size) was likely to have four or five physicians, one clergyperson, one hospital administrator, and sometimes an attorney or nurse. As committees enlarge, they tend to add friends of those already on the committee and people who can be expected to toady to hospital policies (Tonti-Filippini 1996, 339).

Who can join bioethics associations or attend open conferences? Those with sufficient funds. Who gets invited to closed conferences? Those who know the right people. Who can be called a "bioethicist"? Any M.D. who wants to be. But attorneys, clergy, persons with master's degrees in bioethics, and philosophy doctorates—they must have connections.[4]

The field has come to give an imprimatur to present-day medical practice. The endorsement of an ethics consultant or committee can and often does serve to legitimate the status quo. As Rebecca Dresser puts it, we bioethicists have an "allegiance . . . to the people whose very invitation is a prerequisite to our participation" (1996, 156).[5] Indeed, she claims, "It may be that in many of today's health care settings, the bioethicist poses even more danger to patients than physicians do" (ibid.). Benjamin Freedman asks, "What if the physician is pleased with the consultant because he or she has provided an ethical *imprimatur* on behalf of the corrupt course of action that was planned?" (1994, 111) and Nicholas Tonti-Filippini decries the "health care practitioner who . . . [after] attending short bioethics courses, often tends to use the smattering of the language of ethics gained only to validate a lifetime of medical malpractice" (1996, 335).

Bioethics basks in the cognitive and social authority of medicine. *"Cognitive authority* means the authority to have one's description of the world taken seriously, believed, or accepted generally as the truth" (Wendell 1996, 117).[6] Medicine's *social authority* ensues because society automatically accepts medicine's pronouncements, descriptions that then determine how social institutions (such as courts, schools, insurance companies) control our lives (Sherwin 1992, 191; Wendell 1996, 117).[7] These two types of authority grew in the first two-thirds of the twentieth century: as medicine became our monarchy, doctors became our kings. Philosophers who yearned to become philosopher-kings

could go into bioethics. However, as this century draws to a close, executives and administrators have usurped this authority and the monarchy: complaints about their huge salaries and inferior delivery of
health care go largely unnoticed, perhaps because they are under the
protection of the cognitive/social authority of medicine.

Why are health consumers and lower-level health workers steadily
getting less and less while bioethicists get more and more? Is this
merely a coincidence? Is bioethics *complicit* in any way? Is it only indifference (Miles 1997), or are bioethicists simply fortunate to find themselves inside the same cocoon as physicians in this cold war/post–cold
war period? Have they been helplessly swept by the tide of global capitalism, which has widened all gaps between rich and poor? And why
does bioethics tolerate this (Purdy 1996, 2, 3)? Why has the field done
so little to stop the downgrading of health care for those very persons
whose lot an applied ethics ought to improve?

This neglect is strange, because from the inception of the field, "bioethicists have fought to protect vulnerable patients and research subjects from harm and to establish their moral and legal rights" (Wolf
1996b, 6, 10, 13). Yet all this time they were generally blind to gender-
related inequities (Purdy 1996, 7). Currently, to be sure, some bioethicists do point to the gap in health care services as an important ethical
issue to tackle. "Access" seems to be the acceptable buzzword: in a few
sessions in major bioethics conferences, that word appears on the program. One concerned bioethicist, Steven Miles (1997, 2), reports, however, that only two out of fifty articles in the 1996 *Hastings Center Report*
addressed "access," and he claims that "the lack of universal health insurance is the major and distinguishing moral failure of the U.S. health
care system" (1997, 1). Hilde Lindemann Nelson and James Lindemann Nelson call American health care "conspicuously unjust" (1996,
351). Yet I continue to be struck by the remarkable coincidence of bioethics' rise in prestige with the increased disenfranchisement of marginal health care consumers.

As bioethics has grown, women have been involved, playing "an important part in the field from the start" (Wolf 1996b, 5), but that "part"
is peculiar in at least two ways. The growth period of bioethics coin-
cided with the "second wave" of feminism; therefore, certain women
were allowed to play. In that concurrent societal ethos, women were
being permitted more play space in the fields feeding into bioethics:
medicine, philosophy, and law. In this era of "token-plus," instead of
one woman speaking at the conference or contributing to the anthology, organizers invited two.[8] In published papers, the generic "she"
has replaced almost everywhere the generic "he" in referring to a physician. However, the young women entering bioethics were not the

feminists who fought for women's entry; these women, once their entry was granted, had to focus on continuing employment. With the exception of a few feminist philosophers and lawyers (generally Feminist Approaches to Bioethics members), they tackled the same topics in the same ways the boys did. One cannot usually distinguish papers written by such women from those by men. In fact, first papers written by women often parroted the principles (sometimes called the "Georgetown mantra") promulgated in the most popular bioethics text, Thomas Beauchamp and James Childress's *Principles of Biomedical Ethics*, now in its fourth edition (1994). To "add women" to this mantra "and stir" has not improved the recipe and is not going to heal bioethics.

Not only have women's contributions in bioethics been largely indistinguishable from men's, but also peculiar is the fact that the field has essentially ignored academic work using gender and sex as analytical categories (Purdy 1992, 12; Wolf 1996b, 8). Even though bioethics deals with patients in power relationships, and even though its most divisive issues deal with events within women's bodies, its analyses rarely consider the gender inequalities created or reinforced by proposed solutions (Purdy 1996, 4, 7). It shows little concern for ethical treatment of women health care professionals; it ignores the women's health movement's powerful critique of health care; and it fails to make use of the very appropriate methodology developed in feminist scholarship (Wolf 1996b, 10–14).

The percentage of women in medicine increased during this bioethics growth period. But, according to Dresser (1996, 145), "the sexism and sheer brutality of their training programs leaves many of them as detached and unhelpful as their male counterparts tend to be." There's a close parallel in bioethics. Although there may not be "sheer" brutality in the field of bioethics, a kind of subtle brutality conditions female practicing bioethicists to revere authority figures in medicine and to dampen any critical views of the assumptions underlying mainstream bioethics discourse and standard medical practice. To keep a job they may have to compromise their personal integrity (Freedman 1994, 123–128).

I wish to describe two cases of bioethics in action, one demonstrating what it takes to advance a career in bioethics, the other revealing the disempowerment of patients despite bioethics' long-standing concern about vulnerable patients. According to George Annas (1997, 80), "Bioethicists have acted as enablers to permit professional self-interest and institutional profits to trump the interests of patients and their caregivers."

My first case and one of Annas's prime examples is bioethicists' justification of the "Pittsburgh Protocol for Procuring Organs from Non-Heart-Beating Cadavers." This protocol is a scheme for patients or families to "permit life support equipment to be discontinued in an operating room so that [after two minutes with no heartbeat, instead of the standard six minutes] . . . the heart, liver, and kidneys [can be] harvested immediately" (Annas 1997, 78). In the June 1993 issue of the *Kennedy Institute of Ethics Journal*, thirteen of fifteen bioethicists argue that here the end justifies the means. Yet there's no argument that procuring more organs and thus expanding the number of candidates for transplants benefits "either society or the transplant recipients as a whole" (ibid., 79). Currently in the United States, efforts are made to use all organs donated; therefore, the available supply limits the number of transplants performed. More organs available would increase the public cost (heart transplants in 1986 cost $60,000 to $100,000 each) and provide organs to sicker recipients less likely to benefit from them (Hansmann 1989, 79; Annas 1997, 79). Renée Fox, one of the dissenting minority among these authors, argues that the whole procedure is inhumane, for the most dreadful part is "high-tech" death as patient-donors die beneath operating room lights, their bodies already prepared for "the eviscerating surgery" (Fox 1993, 236). Fox also points out that, although these journal articles advance the careers of the commentators (as well as the transplant surgeons), the procedure contaminates the dying process for the patient, the family, and the caretakers (Annas 1997, 79; Fox 1993).

My second example is the case of Baby K, an anencephalic baby born with most of her brain missing. Diagnosed in utero, she was born prematurely by cesarean section in Fairfax Hospital, Virginia, in October 1992 to an African American woman. She was put on a ventilator since the mother (whom I'll call "Mom K") insisted that "everything possible" be done to keep the baby alive. As Mom K told the story at a bioethics meeting in Pittsburgh in 1994, after a few days she was brought before the ethics committee, making it sound like a naughty schoolgirl being sent to the principal's office. Before this, nobody from the committee had tried to hear her point of view or the personal factors that made her want to hang on to this beloved scrap of life. The hospital called in the committee to enforce the policy already decided: that any treatment of Baby K was "medically inappropriate." When Mom K stood firm, the hospital petitioned the federal court. Mom K's attorney won the case there and again later in the Court of Appeals for the Fourth Circuit in February 1994 (Annas 1995; Paris et al. 1995).[9] Baby K, permanently unconscious but with a few reflexes, was weaned from the ventilator and moved to a nursing home; over a period of two and one-half years, she was sent back to the hospital whenever she was in respiratory distress, until she died of pneumonia and heart failure (Paris et al. 1995). Most women would not want to spend two and one-half years of their lives visiting such a being in a nursing home; almost everyone would deplore this waste of resources and public monies—but that sidesteps my point. My point is that the ethics committee did nothing more than represent the cognitive authority of the institution.

Unfortunately, ethics committees and consultants may be forced to neglect the interests of patients and family members. Freedman is especially concerned about this neglect: "[An] ethical consultant must be prepared for the possibility of losing his or her job at any time over some action needed to protect the interest of patients" (1994, 113). In many situations, the consultant can be called to a case only if the attending physician makes that request and cannot talk to the patient and/or family without that same physician's permission (ibid., 119), thus the patient can rarely initiate a consultation (Youngner et al. 1983). "Ethics committees may even proclaim their mission to be protection of patients' rights. Yet at the same time the committee may accord no procedural protections at all, not even notice and a chance to participate in the committee's proceedings" (Wolf 1991, 823). Susan Wolf calls it only a myth that "committees are good for patients" (ibid., 811).

Some who claim that bioethics consultants and committees have become too powerful may be identifying the wrong scapegoat.[10] Bioethicists have essentially no power if they challenge the status quo. What are too powerful are established medicine and those who control it financially; if their spokespersons are bioethicists, then these may appear powerful to the media. One example would be the reports of the Ethics Committee of the American Society for Reproductive Medicine (formerly the American Fertility Society) (American Fertility Society 1994). Ethicists on that committee, including a Catholic priest, are enthusiasts for the latest innovations in reproductive technology and offer few challenges to any new development.

The Far Side of the Gap

Who are the persons on the far side of the gap, who are marginalized in the U.S. health care system and whose condition worries me? Four particular groups are relevant here.

The Sick

All those who are sick or disabled are vulnerable; the gap widens between them and all healthy persons, but especially between them and those who have the power, or think they have the power, to heal them. Being sick or disabled is analogous to "sinning" in our system, for we act as though good health or an unimpaired body were a moral value. Consciously or unconsciously, we blame the victim: he didn't wear a seat belt; she smoked; he ate the wrong kind of mushrooms; she let herself catch German measles. Physicians who have become patients in their own hospitals have felt this disempowerment. In her superb analysis of society's attitudes toward people with dysfunctional bodies, Susan Wendell (1996, 96) points out that the chronically ill are even further marginalized because doctors cannot "control" their illnesses.

"Others"

Nondominant, powerless groups in any society are, by definition, always marginalized. In the United States, these comprise non-Euro-Americans, nonwhites, the poor, homosexuals, and people with little education. When persons in these groups become sick, they are doubly marginalized. Some have the fear (sometimes quite rationally) that their illnesses may be caused by those holding power in society (Dula 1994; Gamble 1997). They may hesitate to approach the medical establishment because they anticipate being treated with contempt (Sherwin 1996, 61) or being given an ineffective or even dangerous "cure" (Gamble 1997). This fear exacerbates their marginalization.

Lower-Echelon Health Care Workers

Huge differentials exist in salary, power, and prestige among health care workers. CEOs of health maintenance organizations and of health insurance firms are among the most highly paid executives in the United States (Freudenheim 1995; Mathews 1995). Many physicians, especially brain surgeons, occupy one side of a gap across from nurses, orderlies, and technicians; and these, in turn, are across a gap from janitors, nurse's aides, and kitchen staff. The salary gaps are unconscionable. Low-paid workers often helplessly witness, firsthand, the negative results of decisions approved by ethics consultants.

Women

Between women and men, not only are the intergroup differentials greater, but gaps also exist within groups. Sick women are offered less aggressive treatment than men, and symptoms they report are more often dismissed or discounted. Women physicians have to put up with sexist medical education and sexual harassment. Among the low-status, low-paying health care positions, the majority of workers are women (Boston Women's Health Book Collective 1992, 652; Dresser 1996, 145, 149). Another group, female family members at home—those who care for patients discharged too soon from hospitals—is so marginalized as to be invisible to the health care system (Nelson and Nelson 1996, 357; Warren 1992, 33). They earn nothing, while early discharge policies allow CEOs to expand their annual salary increases.

Why Does Bioethics Fiddle?

Why do these gaps exist? Why does bioethics fiddle while health care burns? Are these gaps inevitable with the current expansion of global capitalism, so that there is nothing an applied ethics can do to close them? Does bioethics' alliance with institutionalized power keep it closed to issues that concern marginal groups? Wolf points to what she calls "the deep structure of bioethics" to answer the question she poses, "Why has bioethics paid so little attention to gender and to feminist work?" (Wolf 1996b, 5). I am grateful to have the help of her insights and analyses for ferreting out answers to my questions.

Abstraction

Bioethics from the start has drawn on principle-based reasoning and on deductions from principles: usually the principles of autonomy, nonmaleficence, beneficence, and justice (Beauchamp and Childress 1994, 38). This approach has been gathering its nonfeminist critics.[11] Some of those critics and many feminists challenge "the requirements of universality and impartiality" that overlook "the importance of partiality, context, and relational bonds in moral life" (Wolf 1996b, 15), but abstraction still dominates. Furthermore, although marginalized groups have unique standpoints, those views, as nonuniversal, get eliminated from consideration. In fact, many ethics discussions seem oblivious to such groups' very existence: rarely does anyone get the chance to say, "As an African American, how do I look at this issue?"

Individualism and Autonomy

One of the grievous faults Wolf finds in the deep structure of bioethics is its "embrace of liberal individualism that obscured the importance of groups" (1996b, 14). As Fox and Judith Swazey put it, "It is the individual, seen as an autonomous, self-determining entity, rather than in relationship to significant others, that is the starting point and the foundation stone of American bioethics" (1984, 339). To be sure, feminists have struggled for women's right to self-determination; we want each and every woman to be treated as a full person. But if we look only at individual rights, then we may not see the groups to which such individuals belong. These groups are important and necessary for self-esteem, physical and emotional survival, and the joys and pleasures in life. But certain groups are marginalized, so that any member of that group, so labeled, also gets marginalized. If we focus on individual rights, groups remain marginalized; in fact, such approaches in bioethics may exacerbate group marginalization.

The Reagan–Bush approach that made a virtue of individualism (Hunt 1996, 1) created even more marginalized groups. Leonard Harris argues that "autonomy . . . can be a dubious good for persons socially constructed as inferior" (1992, 144), because "the concept of autonomy readily allows health care professionals to blame victims, render less than adequate services, or withhold cures" (143). According to Wolf, "mechanical application of the rights equation . . . will wrongly assume that all [persons have] the resources of the idealized rights-bearer—a person of means untroubled by oppression" (1996a, 299).[12]

Social Movements

Bioethics nestles happily inside its little cocoon. It seems almost unaware of social movements such as the women's movement, especially the women's health movement and the midwifery debate, and, in Wolf's view, of such academic developments as critical race theory (Wolf 1996b, 16). Criticisms by maverick males of the assumptions underlying established medicine have been all but ignored, such as those by Rick Carlson, Michel Foucault, Charles Inlander, Robert Mendelsohn, Thomas Szasz, and Ivan Illich. These social movements, academic trends, and criticisms of established medicine plead the case of marginalized groups; many of them focus specifically on demarginalization, in both theory and activism. They advocate elimination of differential treatment as part of the demarginalization process. Sometimes laypersons also question the cognitive authority of medicine. By ignoring both the movements and the questioning, and by ignoring the fiscally homeless in health care, bioethics stays in an ivory tower (Miles 1997) while reinforcing in concrete the marginalized position.

Unexamined Assumptions

Bioethics doesn't "searchingly identify and evaluate the presuppositions and assumptions on which it rests" (Fox and Swazey 1984, 356).

One such assumption may be that health is a virtue: this leads to blaming the sick. Bioethics may debate how much aggressive treatment to give to a compromised newborn but not what can be done to prevent so many premature births. It may ask which candidate should get a donor heart but not whether parts from other humans should be used to extend lives. And a 1996 request for proposals put out by the Ethical, Legal, and Social Implications Branch of the Center for Human Genome Research (U.S. National Institutes of Health) asked for proposals on *how* to carry out the Human Genome Diversity Project (collecting DNA samples from ethnic groups around the world) but not on *whether* to carry out the project. Many members of a truly marginalized group in the United States, Native Americans, are among those who object to this project's very existence (Crigger 1995). Ignoring their objection further marginalized them.

Since bioethicists are usually unaware that they work from implicit assumptions, many deliberations and theoretical issues that excite them are not shared by sick, disabled, and poor persons. For example, problems arise with the doctrine of informed consent, supposedly devised to protect patients. According to anthropologist Pamela Sankar, "Perhaps informed consent practices and language reflect so strongly the tacit understandings—the culture—of bio-medicine or of bioethics that [their] intent is opaque to others" (1997, 2). This flaw permeates bioethics discourse.

Crisis Orientation

Bioethics seems to focus on medical emergencies, such as whether to continue life support; or it gets mesmerized by new technologies or new potential techniques, such as egg donation or cloning, which at most will affect only a tiny percentage of patients. "The allocation of nonmaterial resources such as personnel, talent, skill, time, energy, caring, and compassion is rarely mentioned" (Fox and Swazey 1984, 353). Virginia Warren, to whom the emphasis on "crisis issues" is a major flaw in mainstream bioethics, urges that, to work toward more equitable health care, bioethicists ought to focus more on "housekeeping issues" (1992, 36–38). Among these issues she includes job-related stress in health care professionals, with concomitant alcoholism, drug abuse, and high divorce rates (ibid., 36); and she asks how to "foster the conditions which make informed consent more likely" (ibid., 48).

Arthur Caplan has described how well a hospital handled crisis issues as his father-in-law died from colon cancer. For example, they accepted his wish for no extraordinary forms of life support. However, what Caplan calls "microethical problems" (and Warren would call "housekeeping issues") made the hospital experience vexing for both patient and family members: "delivery of three meals a day for weeks to a man who could not possibly eat a morsel . . . a continuing parade of persons . . . in the patient's room . . . 24 entries in one shift . . . lights and noise [that create] a perfect sleep-free zone" (Caplan 1995, 42). Despite handling the actual crisis satisfactorily, the hospital still managed to marginalize the patient and his family.

False Dualisms

Bioethics texts for students are filled with case studies in crisis issues presented as dilemmas. Which principle should trump? Terminate pregnancy or not? Resuscitate or not? Does the person in the coma still believe in the consent form she signed last year: yes or no? "The representation of moral problems as . . . forced choices between two (or more) equally unwelcome alternatives . . . is destructive . . . and discourages the attempt to devise better alternatives" (Whitbeck 1993, 3). Fox and Swazey deplore "dichotomous distinctions" and "bipolar choices" (1984, 355). As presented by Carol Gilligan (1982), Lawrence Kohlberg's "Heinz dilemma" clearly illustrates this: in this scenario, Amy refuses to view the problem presented to her as a lose-lose dilemma. She chooses neither to steal a drug nor to let Heinz's wife die, but instead uses her creativity and sensitivity to suggest several win-win solutions.[13]

And according to Cheryl Sanders, one reason African Americans may not be welcome in bioethics discourse, or may themselves feel uncomfortable there, is that "the African American ethos is holistic and nondualistic, emphasizing that most matters are better understood in terms of 'both-and' rather than 'either-or' " (1992, 165). Constructing problems as dilemmas may also exacerbate power conflicts over who has the moral authority to make the final decision (Warren 1992, 38–39). And, as in the Baby K case, dichotomous thinking precipitates legal intervention (Annas 1995).

Baby K Revisited

Let's reconstruct the case of Baby K to alleviate these shortcomings of bioethics. The hospital staged it as a conflict of principles between Mom K's autonomy and the beneficence/paternalism of the doctors who knew it would do Baby K no good to prolong her unconscious life. Neither side would give in, so this crisis issue had to go to court to be settled (Annas 1995; Paris et al. 1995). Now let's call in Amy as a consultant: she suggests we start at the housekeeping level. The hospital staff knew well ahead of the birth that Baby K would be anencephalic; indeed, they had urged Mom K to get an abortion. Once Amy learned that Mom K had unrealistic hopes that God would cure her child, she recruited a sympathetic listener and then let Baby K be born vaginally, not by cesarean section. Cesarean sections may be ordered to save compromised babies, to give practice to residents, or to get larger payments from Medicaid; none of these reasons should hold here. With a vaginal birth, Baby K would most likely have been stillborn, and then Mom K could have held her and started her grieving process. She would not have had the stress of neonatal death compounded by recovery from surgery. Amy would arrange for Mom K to use the hospital chapel for grieving and, later, for a memorial service. Amy would then set up appointments with the hospital nutritionist and provide Mom K with folic acid supplements during her next pregnancy.[14]

Closing the Gaps

What can feminist bioethicists do to close these gaps? Unfortunately, we can do very little. Those of us who wish to get tenure and to have papers published—that is, to get paid for "doing" ethics—can write and speak only within accepted paradigms.[15] We can't rock boats. The deep structure of bioethics needs to be changed, as Wolf has indicated. And this is unlikely to occur, because so many people *enjoy* the status quo. It's such intellectual fun to debate issues like futility, cloning, and procuring organs for transplant. And bioethicists want to keep a share of the prestige that doctors have in our society.

But things have gotten about as bad as they can get for the marginalized. So what can *we* do? By "we," I mean people of goodwill who are genuinely concerned about the marginalized but who have no real power in either medicine or bioethics.

Attention to Children

Perhaps as women we may be heard if we argue on behalf of children. If children (of all groups) can obtain proper attention in the health care system and in the basics of food and shelter that are essential to health, then—since children grow up—maybe those same individuals will someday continue to see that others are provided for. In summer 1997, Congress passed a balanced-budget bill that will extend health care coverage to uninsured children (Balanced-budget bill 1997).

Even though this particular plan uses the existing insurance industry and thus further lines its pockets (Finkelstein 1997), it will likely shrink some gaps. According to Mary Hunt (1996, 3), "The common good ahead begins with the survival of the children on the bottom . . . as a first step toward reorganizing society."

Health Insurance, Employment, and Health Outcomes

We may have some clout in advocating that health insurance *not* be tied to employment. Then the poor, the unskilled, and the disabled who cannot get jobs would have access to health care. Currently, employers may avoid hiring people who, they think, are likely to require expensive treatment.

We can try to initiate a focus on *outcomes*, possibly a way for concerned bioethicists to scratch at the edifice of medicine. By "outcomes," I mean whatever actually leads to better health for more people. For example, meals for pregnant women and access to drug-treatment programs lead to better outcomes than inventing ways to keep one-pound babies alive in the intensive care nursery. Here again is a fundamental, housekeeping, crisis-preventing solution to "set health care delivery into the broader social context" (Nelson and Nelson 1996, 366). Work on these issues, rather than on crises, will accomplish more demarginalization.

Alliances with Concerned Physicians

Many physicians find that their work as healers is seriously compromised by the proliferation of managed care. They are concerned about the "alliance of economic interest between service providers and upper-income citizens" in health financing (Evans 1997). Physicians for a National Health Program is a key group;[16] several of its members have documented the gaps in health care with figures and statistics, for example, the extent of deterioration in service provided and the excess costs of health administration (Woolhandler and Himmelstein 1997). Just as we need the support of physicians to keep our jobs as consultants and stay on ethics committees, they may well find us useful allies in fighting corporate medicine.

Guidance from the Women's Health Movement

We should look carefully in the publications of the women's health movement to find actions to take (Dresser 1996). They have succeeded in exposing mistreatment of women and in fostering improvements, through such means as simple persistence, press conferences, writings, and modeling of appropriate patient–provider behavior. Although the movement started as a middle-class phenomenon, it realized, right from the start, that marginalized groups are treated worse than the women who became the early activists. In order to act in the best interests of groups we hope to demarginalize, we need those very groups' active participation in determining the directions to take. The women's health movement has an outstanding record of finding, fostering, and inspiring members of disadvantaged groups to get involved in policy making.

Dresser urges feminist bioethicists in hospitals to pursue a variety of actions, especially regarding our own behavior. Her long activist list should be posted at the desk of every feminist in clinical ethics. Examples from this list include supporting female medical students, respecting the work of underpaid and undervalued health care workers, not adopting the problematic habits of medical colleagues, and including women's health movement literature in the courses we teach (Dresser 1996, 152–157).

Conclusion

In 1992, Laura Purdy and I issued "calls" to heal medicine and to heal ethics (Holmes 1992; Purdy 1992). What has happened since then? We could point to medicine's excellent health, with CEO salaries much healthier than in 1992. We could point to bioethics' excellent health, with its increased sharing of medicine's salaries and prestige. We could simply not notice the continuing expansion of the gaps between the haves and the have-nots. But, no, let's remove those blindfolds. We must realize that both medicine and bioethics still need to be healed.

Notes

1. Indeed, the problems resulting from shortened hospital stays became extreme enough for even our callous Congress to take notice in 1997: new mothers and women with mastectomies are now permitted—by law—forty-eight hours in the hospital.

2. Some analysts put the blame for this discrepancy on stockholders who prefer hospitals that put profits over quality care and support high salaries for CEOs who can bring profits to investors.

3. I use the term "bioethics" instead of "medical ethics" or "biomedical ethics" because that term has been chosen for this book and in general is more commonly used. In this chapter, however, I discuss only medical issues.

4. Fox and Swazey are concerned about the small number of social scientists working in bioethics and believe that the profession suffers without their insights (1984, 350).

5. Carl Elliott (1997) raises a larger question, whether moral advice should be bought and sold: "Who chooses the ethicist and why? . . . [T]here is generally no shortage of ethicists who will support any number of moral stances on a complex issue."

6. For an insightful analysis of the insidiousness of the cognitive and social authority of medicine, especially of how it can validate or invalidate persons' perceptions of their bodies, see Wendell (1996, 117–138).

7. The term "medicalization," used by some authors, may refer to the social acceptance of the cognitive authority of medicine. See, for example, Dresser (1996, 148).

8. See also Sherwin (1996, 59–60).

9. The courts claimed that Baby K was protected by the Emergency Medical Treatment and Active Labor Act of 1985, the Rehabilitation Act of 1973, and the Americans with Disabilities Act of 1990 (Annas 1995; Paris et al. 1995). However, it is unusual that Mom K was able to get judicial review. More commonly, as Susan Wolf reports, families will find that an ethics committee's support for a physician's recommendation is "a nearly insuperable obstacle" or that "limitations on the family's resources may make the committee's advice the last word." Therefore, "the theoretical availability of judicial review is of no practical help" (Wolf 1991, 809).

10. See Arthur Caplan's comment on the complaint that bioethics consultants and committees have become too powerful (1997, 5).

11. For lucid and perceptive analyses of the "principlism" of Beauchamp and Childress (1994), nonfeminist criticisms of it, and Beauchamp and Childress's responses to their critics, see Tong (1997, 62–70). See also Fox and Swazey's discussion of reductionism (1984, 357) and Walker's (1989) examination of universalism.

12. In the 1990s, autonomy bashing and rights bashing by nonfeminists and feminists alike have become quite popular. Some of this work is excellent, although feminists need to examine it carefully. Some principlists complain that other principlists use autonomy as trumps, whereas beneficence really should trump autonomy. Also, some nonfeminists bash autonomy because they are antichoice and don't want women to have the autonomy to "choose" abortion. Without a doubt, women need the option of abortion, and pregnant women in labor need the autonomy to have their choices respected about the method of delivery. For better sources on the new "deconstruction" of autonomy, see Wolf (1996a, 314–315 nn 70, 75, 78); Sherwin's chapters on autonomy/paternalism (1992, 137–157) and on a relational approach to autonomy (1998, 19–47); Donchin (1998); and especially Smith (1997)

13. Most authors analyze the Heinz dilemma as a conflict between the principles of justice and caring. Amy, however, does not reason about which principle should trump, but instead about finding a solution that takes all pertinent values into account. See also Purdy's utilitarian analysis (1996, 13).

14. But suppose that there is a legitimate reason for the cesarean or that Mom K has asked for one. In that case, Amy would see that Mom K has an anesthetic that will allow her to witness the birth, get to hold the baby right away, and participate in what will turn out to be fruitless attempts at resuscitation. Nobody would be belligerent and argue; the term "medically inappropriate" would not be bandied around. Mom K would hear such things as, "What a sweet face! Lovely little hands and feet!"

15. I have taken a personal stance not to join any bioethics organizations because of their elitism. I am, however, an active member of the International Network on Feminist Approaches to Bioethics because of its strong nonelitist stance and plan to continue to participate as long as it says true to this position.

16. Physicians for a National Health Program, 332 South Michigan, Suite 500, Chicago, IL 60604; telephone: (312) 554-0382.

References

American Fertility Society, Ethics Committee. 1994. Ethical considerations of assisted reproductive technologies. *Fertility and Sterility* 62 (5, suppl.1): iii–125s.

Annas, George J. 1995. Asking the courts to set the standard of emergency care: The case of Baby K. *New England Journal of Medicine* 330 (21): 1542–1545.

———. 1997. Review of *Procuring Organs for Transplant: The Debate over Non-Heart-Beating Cadaver Protocols*, (ed. Robert M. Arnold, Stuart J. Youngner, Renie Schapiro, and Carol Mason Spicer. [Baltimore: Johns Hopkins University Press, 1995]) *Bioethics* 11 (1): 77–80.

Balanced-budget bill extends health care to children. 1997. *Nation's Health*, August; 1, 5.

Beauchamp, Thomas L., and James F. Childress. 1994. *Principles of Biomedical Ethics* 4th ed. New York: Oxford University Press.

Boston Women's Health Book Collective. 1992. *The New Our Bodies, Ourselves: A Book by and for Women*. New York: Simon & Schuster.

Caplan, Arthur. 1995. *Moral Matters: Ethical Issues in Medicine and the Life Sciences*. New York: Wiley & Sons.

———. 1997. From the director: Why don't they love us anymore? *Center for Bioethics Newsletter* (University of Pennsylvania) 2 (3): 1, 5.

Cohen, Cynthia. 1988. Birth of a network. *Hastings Center Report* 18 (1): 11.

Crigger, Bette-Jane. 1995. The "vampire project." *Hastings Center Report* 25 (1): 2.

Donchin, Anne. 1998. Understanding autonomy relationally: Toward a reconfiguration of bioethical principles. *Journal of Medicine and Philosophy* 23 (4): in press.

Dresser, Rebecca. 1996. What bioethics can learn from the women's health movement. In Susan M. Wolf, ed. *Feminism and Bioethics: Beyond Reproduction*. New York: Oxford University Press.

Dula, Annette. 1994. African American suspicion of the healthcare system is justified: What do we do about it? *Cambridge Quarterly of Healthcare Ethics* 3: 347–357.

Elliott, Carl. 1997. Bioethics as commodity: Does the exchange of money alter the nature of an ethics consultation? *Bioethics Examiner* 1 (3): 1–2.

Evans, Robert G. 1997. Going for the gold: The redistributive agenda behind market-based health care reform. *Journal of Health Politics, Policy & Law* 22 (2): 427–465.

Finkelstein, Katherine Eban. 1997. Insuring children: Health care reform writ small. *Nation* March 3: 18–21.

Fox, Renée C. 1993. An ignoble form of cannibalism: Reflections on the Pittsburgh Protocol for Procuring Organs from Non-Heart-Beating Cadavers. *Kennedy Institute of Ethics Journal* 3 (2): 231–240.

Fox, Renée C., and Judith P. Swazey. 1984. Medical morality is not bioethics: Medical ethics in China and the United States. *Perspectives in Biology and Medicine* 27 (3): 336–360; also published in Renée Fox, ed. *Essays in Medical Sociology*. New Brunswick, NJ: Transaction Books, 1980.

Freedman, Benjamin. 1994. From avocation to vocation: Working conditions for clinical health care ethics consultants. In Françoise Baylis, ed., *The Health Care Ethics Consultant*. Totowa, NJ: Humana Press.

Freudenheim, Milt. 1995. Penny-pinching H.M.O.'s showed their generosity in executive paychecks. *New York Times,* June 15: D1, D4.

Gamble, Vanessa Northington. 1997. Under the shadow of Tuskegee: African Americans and health care. *American Journal of Public Health* 87 (11): 1773–1778.

Gilligan, Carol. 1982. *In a Different Voice: Psychological Theory and Women's Development.* Cambridge, MA: Harvard University Press.

Hansmann, Henry. 1989. The economics and ethics of markets for human organs. *Journal of Health Politics, Policy & Law* 14 (1): 57–85.

Harris, Leonard. 1992. Autonomy under duress. In Harley E. Flack and Edmund D. Pellegrino, eds. *African-American Perspectives on Biomedical Ethics.* Washington, DC: Georgetown University Press.

Holmes, Helen Bequaert. 1992. A call to heal medicine. In Helen Bequaert Holmes and Laura M. Purdy, eds., *Feminist Perspectives in Medical Ethics.* Bloomington: Indiana University Press.

Hunt, Mary E. 1996. Whither the common good. *Waterwheel* 9 (3): 1–3.

Mathews, Jay. 1995. $6.1 million a year to run an HMO. *New York Times*, December 27: B1–B2.

Miles, Steven. 1997. The role of bioethics and access to US health care: Is bioethics one of Kitty Genovese's neighbors? *Bioethics Examiner* 1 (2): 1–2

Nelson, Hilde Lindemann, and James Lindemann Nelson. 1996. Justice in the allocation of health care resources: A feminist account. In Susan M. Wolf, ed., *Feminism and Bioethics: Beyond Reproduction.* New York: Oxford University Press.

Paris, John J., Steven H. Miles, Arthur Kohman, and Frank Reardon. 1995. Guidelines on the care of anencephalic infants: A response to Baby K. *Journal of Perinatology* 15 (4): 318–324.

Purdy, Laura M. 1992. A call to heal ethics. In Helen Bequaert Holmes and Laura M. Purdy, eds., *Feminist Perspectives in Medical Ethics*. Bloomington: Indiana University Press.

———. 1996. *Reproducing Persons: Issues in Feminist Bioethics.* Ithaca, NY: Cornell University Press.

Sanders, Cheryl J. 1992. Problems and limitations of an African-American perspective in biomedical ethics: A theological view. In Harley E. Flack and Edmund D. Pellegrino, eds., *African-American Perspectives on Biomedical Ethics.* Washington, DC: Georgetown University Press.

Sankar, Pamela. 1997. Anthropology in bioethics. *Center for Bioethics Newsletter* 2 (2): 1–2.

Sherwin, Susan. 1992. *No Longer Patient: Feminist Ethics and Health Care.* Philadelphia: Temple University Press.

———. 1996. Feminism and bioethics. In Susan M. Wolf, ed., *Feminism and Bioethics: Beyond Reproduction.* New York: Oxford University Press.

———. 1998. A relational account of autonomy in health care. In Susan Sherwin, ed. *The Politics of Women's Health: Exploring Agency and Autonomy.* Philadelphia: Temple University Press.

Smith, Janet E. 1997. The pre-eminence of autonomy in bioethics. In David S. Oderberg and Jacqueline A. Laing, eds., *Human Lives: Critical Essays on Consequentialist Bioethics.* New York: St. Martin's Press.

Tong, Rosemarie. 1997. *Feminist Approaches to Bioethics: Theoretical Reflections and Practical Applications.* Boulder, CO: Westview Press.

Tonti-Filippini, Nicholas. 1996. Review of *The Health Care Ethics Consultant,* (ed. Françoise Baylis [Totowa NJ: Humana Press]) *Bioethics* 10: (4) 334–340.

Walker, Margaret Urban. 1989. Moral understandings: Alternative "epistemology" for a feminist ethics. *Hypatia* 4 (2): 15–28.

Warren, Virginia L. 1992. Feminist directions in medical ethics. In Helen Bequaert Holmes and Laura M. Purdy, eds., *Feminist Perspectives in Medical Ethics.* Bloomington: Indiana University Press.

Wendell, Susan. 1996. *The Rejected Body: Feminist Philosophical Reflections on Disability.* New York: Routledge.

Whitbeck, Caroline. 1993. The trouble with dilemmas: Rethinking applied ethics. *Newsletter of the Network on Feminist Approaches to Bioethics* 1 (1): 3.

Williams, Erin. 1997. Managed care and disempowered persons. *Newsletter of the Network on Feminist Approaches to Bioethics* 5 (2): 3.

Wolf, Susan M. 1991. Ethics committees and due process: Nesting rights in a community of caring. *Maryland Law Review* 50: 798–858.

———. 1996a. Gender, feminism, and death: Physician-assisted suicide and euthanasia. In Susan M. Wolf, ed., *Feminism and Bioethics: Beyond Reproduction.* New York: Oxford University Press.

———. 1996b. Introduction: Gender and feminism in bioethics. In Susan M. Wolf, ed., *Feminism and Bioethics: Beyond Reproduction.* New York: Oxford University Press.

Woolhandler, Steffie, and David U. Himmelstein. 1997. Costs of care and administration at for-profit and other hospitals in the United States. *New England Journal of Medicine* 336: 769–774.

Youngner, Stuart J., David L. Jackson, Claudia Coulton, Barbara W. Juknialis, and Era M. Smith. 1983. A national survey of hospital ethics committees. *Critical Care Medicine* 11 (11): 902–905.

4

Erasing Difference:
Race, Ethnicity, and
Gender in Bioethics

Susan M. Wolf

Bioethics was born of outrage at scandals in which difference figured large. Experimentation on Jews and others in Nazi concentration camps and on African Americans in the Tuskegee syphilis trials were pivotal events. The field has come of age decrying the wrongs of physicians and scientists enacting the prejudices of their day, whether those professionals were extolling racist eugenics or excluding women from research trials. The data has abundantly documented worse health care, doctor-patient relationships, and problems for human subjects in research by race, ethnicity, and gender.

Yet, as I argue below, bioethics has until recently largely ignored the significance of these differences. The field has emphasized that Tuskegee teaches the need for disclosure and consent, with less analysis of the detailed workings of racism in research. The field has similarly construed the Nazi physicians' experimentation on concentration camp victims as a case of human subjects abuse, with less focus on how that experimentation functioned as a part of the attempted extermination of an ethnic and religious group. Bioethics has too often recast these and other events as problems between generic subjects or patients and generic doctors. Issues of difference have frequently been avoided.

Even when the racism, ethnocentrism, and sexism haunting key events have been acknowledged, they have usually been treated as a layer of extra insult added onto the more fundamental harm (depriving the generic research subject of the chance to give genuine consent, for instance). At most, bioethicists have called for treating people of

color as whites are treated, or women as men. What is missing is a bioethics analysis that places race, ethnicity, and gender at the center and
delves into the significance of difference.

That kind of analysis is sorely needed. Differences of race, ethnicity,
and gender all too often seem to occasion unethical behavior in health
care and research. Indeed, medicine and biomedical science have
played a troubling role in the very construction of these differences.
"Race medicine," for example, asserted that race had a scientifically
rigorous meaning, that the races had certain medically ascertainable
characteristics, and that this justified disadvantaging certain races (see,
e.g., Bhopal 1997; Witzig 1996). Given this history, it should be no surprise that the scandals that have catalyzed modern bioethics are permeated by difference. The closer one looks, the more it seems that bioethics at its core should be about these issues of difference. Yet the field
has never placed those issues at the center. It is a glaring omission.
After all, bioethics took hold in the late 1960s and early 1970s as a
movement for patients' rights. It grew up alongside the civil rights and
women's rights movements, borrowing from their rhetoric and momentum. Yet it has never taken seriously the linkages.

Bioethics has been stunted by these failures. And the field has yet to
fully recover. After decades of marginalizing race, ethnicity, and gender, bioethics has recently begun paying greater attention—but not by
moving these categories to their proper place at the center of analysis.
Instead, the field has found new ways to keep them at the margin. Bioethics has held fast to its liberal roots, condemning discrimination and
calling for equal treatment but failing to dig deeper into the literature
and debates about difference.

I trace the history below. I then argue that bioethics fails even now
to rise to the challenges posed by race, ethnicity, and gender. I finally
suggest what rising to that challenge might entail.

Several cautions apply to my analysis. Many have cogently argued
that "race" is a fiction, rather than a biological category with clear
boundaries and meaning (see, e.g., Bhopal 1997; Haney López 1994).
Much the same can be said of ethnicity. I take race and ethnicity to be
not biological truths but cultural artifacts. Too often they are axes of
disadvantage, but they can also be a basis for an individual's chosen
affiliation and identity. Gender, too, I take to be a cultural creation, in
contrast to the biological fact of sex.

Second, in arguing that bioethics has failed to come to terms with
race, ethnicity, and gender, I do not mean to suggest that the categories
are equivalent. Racism, ethnocentrism, and sexism are not the same
thing; their history, dynamic, and meaning differ (Grillo and Wildman
1991). Yet I also do not mean to imply that they are completely sepa-

rate. A great many people, such as women of color, experience the intersection of these sources of disadvantage (Crenshaw 1989). And, as many have noted, race, ethnicity, and gender operate as interlocking systems of disadvantage, reinforcing one another (see, e.g., Sherwin 1992, 222–240). History bears witness to the connections among these phenomena to privilege, ultimately, a very limited group.

Finally, in focusing on race, ethnicity, and gender, I do not mean to ignore the significance of economic status or class.[1] Class is a pervasive source of disadvantage in health care (see, e.g., Guralnik and Leveille 1997), and bioethics has been slow to focus on it. The field has addressed issues of access to health care, for example, far later than issues such as physician truth-telling that assumed access and a preexisting doctor–patient relationship (Wolf 1994a). Yet recent debates on health care reform, rationing, and allocation seem to have forced questions of class onto the bioethics agenda. And heavy reliance in those debates on the work of John Rawls has awarded a place in bioethics analyses to "the worst off" (see, e.g., Daniels 1985, 43–44). Somehow, though, this has not translated into a comparable focus on race, ethnicity, and gender, though many of the economically least well off labor as well under burdens associated with those other factors. I therefore concentrate here on the puzzling and graphic failure in bioethics to come to terms with race, ethnicity, and gender.

The History and Harm of Erasure

Throughout the roughly three decades that comprise the history of modern bioethics, race and ethnicity have permeated the problems that the field has analyzed. As noted above, the Nazi doctors' trial at Nuremberg and the Tuskegee study bear early witness (Annas and Grodin 1992; Caplan 1992; Brandt 1985; Jones 1993; King 1992a). Beyond those events, debates about genetic screening are still haunted by the mandatory sickle-cell screening of African Americans in the early 1970s with little therapeutic benefit and considerable cost in stigmatization and discrimination (see, e.g., Andrews 1997, 902–903; Ikemoto 1997, 943–944). Indeed, the racist use of eugenics throughout the twentieth century is a backdrop to the entire human genome initiative (Kevles 1992). Race and ethnicity have figured prominently in myriad realms, not only in the history of human experimentation and genetics, but in population control, with past racist abuse of sterilization (Banks 1989–1990, 361–363; Roberts 1997, 89–98; Roberts 1996, 124). Moreover, debates about health care delivery have long demanded attention to data showing lesser access to treatment and poorer quality of care by race (Ayanian et al. 1993; Wenneker and Epstein 1989; Whittle et al. 1993). Even the American Medical Association has reported on "Black-White Disparities in Health Care" (Council on Ethical and Judicial Affairs 1990). In domain after domain in bioethics, the centrality of race and ethnicity to a moral analysis seems clear.

The fact that society-wide problems of difference have deeply affected medical and research encounters should not be startling. When doctor and patient or researcher and subject meet, they usually are separated by a gulf marked by multiple differences. Bioethics has claimed to focus on that gulf. After all, bioethics has not been about the doctor and patient who have everything in common, see treatment decisions the same way, and meet as equals. The patient hardly needs autonomy, bioethics' most vaunted principle, when the physician already agrees. Bioethics thus has spoken from its beginning about the patient and doctor who are different people with different values. They often have starkly divergent power in the clinic and beyond. The physician's white coat and the patient's gown have disguised their individual characteristics, but in our heterogeneous nation, they frequently have come from utterly different worlds.

Indeed, the scandals that have motivated bioethics have evidenced the repeated failure of physicians and scientists to negotiate their differences with patients and subjects. In fact, those scandals are about the exploitation of that difference. They show how researchers have often used not the privileged but the powerless for experimentation, how health care professionals have too frequently mistreated the most vulnerable, and how our health care system has failed the least well off. Differences of race and ethnicity are central to this story.

But the literature of bioethics has not made race and ethnicity central categories of analysis. Indeed, until recently, that literature failed to treat those categories as particularly important. Neither "race" nor "ethnicity" is even a heading in the index of the influential *Principles of Biomedical Ethics* (Beauchamp and Childress 1994). Bioethicists instead have talked about harms to the generic patient or research subject, occasionally noting as an afterthought that the difficulties may be even worse when the patient or subject is a person of color. Bioethics has been nearly devoid of analyses that started with the circumstances of a person of color and built an argument from there. Nor has bioethics delved into the history of race and ethnicity in medicine to develop an understanding of how physicians and scientists have treated these categories over time. Had bioethicists done this, surely Tuskegee would be regarded as part of a much older pattern marked by horrors such as Marion Sims's nineteenth-century development of gynecological surgery using unanesthetized slave women, some bought for this purpose

(Barker-Benfield 1976, 101). There is a long history bioethics has yet to analyze (see, e.g., Savitt 1978 and 1982).

Much as bioethics has fled from serious engagement with race and ethnicity, it has also avoided gender until recently. The silence surrounding gender has been less shocking in a sense; the founding moments in modern bioethics have not been about gender in the same stark way that they have been about race and ethnicity. But the importance of gender as a category of analysis should have been seen long before it was. After all, women are statistically the majority of patients and have the majority of patient contacts (Benson and Marano 1994, 120; Horton 1992, 93–94). The history of health care in the United States has been permeated by issues of gender, with a dominance of female health care providers in colonial times, ended by the nineteenth-century ascendancy of physicians who were overwhelmingly males (Starr 1982, 49). There is a tradition of medicalizing normal female functioning and construing it as pathological, be it menstruation, childbirth, or menopause (see, e.g., Ehrenreich and English, 1973 and 1979). More recently, there is data showing less access to some forms of treatment and worse quality of care for women (Council on Ethical and Judicial Affairs 1991). We have seen the historical exclusion of women of childbearing age from drug trials and other studies, so that the male model is misconstrued as universal and therapies are marketed without adequate testing on women (Dresser 1992; Merton 1996). And health systems analysis and reform have to attend to gender, since women are less often privately insured, are disadvantaged by any health insurance system based on employer contributions, and predominate among the elderly (Jecker 1993; Nelson and Nelson 1996).

All of these are strong reasons for bioethics to pay close attention to gender. But until recently, it has not.

I have argued elsewhere that the failure in bioethics to analyze gender, race, and ethnicity has deep roots (Wolf 1996b). The demographics of bioethics may have contributed to the avoidance of race and ethnicity, though the same cannot be said of gender, as women have been well represented in the field from the start. The way bioethics is practiced may be a factor, too, as bioethicists usually answer the questions of professionals and government authorities rather than patients, research subjects, and community members. The fact that bioethics has labored to establish itself within medical schools and health centers may have encouraged an intellectual conservatism and the isolation of bioethics from major trends within the academy toward attention to race, ethnicity, and gender.

The explanation rooted deepest in the very structure of bioethics, however, is the field's early embrace of what we now call a "principlist" approach (see Beauchamp and Childress 1994, 37; Wolf 1994b). Bioethicists have tended to ground moral analysis on a roster of four principles—autonomy, beneficence, nonmaleficence, and justice—that is supposedly applicable to all patients, with no attention to race, ethnicity, or gender. But it turns out that these principles do not provide an effective escape from issues of difference. Some ethnic groups tend to reject the primacy of individual autonomy and this particular roster of values (Blackhall et al. 1995; Carrese and Rhodes 1995). Moreover, members of certain ethnic groups present dilemmas that are almost impossible to resolve using the usual set of four principles. (One example would be the dilemma presented by a Hmong woman rejecting a cesarean section at full term because of negative cultural beliefs about surgery and thus endangering her fetus.) These dilemmas seem to require attention to other values such as respect for cultural affiliation and community context. Finally, it seems that the individualistic patient-by-patient way in which bioethics has applied principles such as autonomy and beneficence has obscured patterns of stereotyping and prejudice that are themselves ethical problems.

The erasure of difference in bioethics has not simply been an artifact of the field's early embrace of principlism. As I have argued elsewhere, it has been a legacy of the field's commitment to a liberal individualism born of the work of Kant and Mill (Wolf 1996c). Bioethics has strained for universals, ignoring the significance of groups and the importance of context. It has rushed to generalize about "the patient," "the subject," "the doctor," and "the researcher." It has claimed to find basic truths about the doctor–patient encounter.

In trying to speak of everyone, bioethics has spoken of no one. There is no such thing as a patient without race, ethnicity, and gender. Bioethical analysis, failing to focus on these differences, thus missed much that was important in real cases: the patient whose candidacy for transplantation is compromised by the racist assumption that he will fail to follow post-transplant treatment instructions, the patient branded "noncompliant" or even "incompetent" because she refuses surgery in keeping with the norms of her ethnic group, the person never offered experimental treatment for her cancer because the approved protocol enrolls only men. A bioethics ignoring race, ethnicity, and gender will fail in moral analysis of these cases.

Beyond the failure to analyze individual cases properly, bioethics thus has failed to see patterns and subject them to ethical analysis. After all, problems of difference in the United States are not just individual blips on the radar. They are systematic patterns of long standing. They have a history that demands study before the full meaning of an individual instance can be understood. And they have complex

manifestations ranging from hate crime and overt racism to stereotyping, unwarranted assumptions, and unconscious bias (see, e.g., Lawrence 1987). Bioethics has left these phenomena largely unanalyzed,
despite their repeated manifestation in medicine and human subjects
research. This, in turn, has left bioethics with an impoverished list of
problems to tackle, dropping racism and sexism in medicine, for example, off the list and dampening interest in correlated problems such as
lack of access to health care and to experimental protocols. It also has
left bioethics a quiet conspirator in the perpetuation of racist, ethno-
centric, and sexist practices (see Purdy 1996, 18). Finally, it has meant
that bioethics is practiced from a narrow perspective (Wolf 1996a).
After all, the only person who might claim that race is of no significance in the United States is a person of the dominant race, asserting
the privilege to ignore it. A person of a stigmatized race would be
hard-pressed to claim race irrelevant in her treatment. The same goes
for ethnicity and gender: only those of dominant ethnicity and gender
would claim those factors to be of no consequence. Thus bioethics has
emanated from the perspective of the dominant race, ethnicity, and
gender. Bioethics has claimed to be unsituated and universal. But it has
been thoroughly situated and quite unreflective about its prevailing
perspective. Only recently have scholars suggested what a different
bioethics, one from an African American or feminist perspective, might
look like (see, e.g., Dula and Goering 1994; Flack and Pellegrino 1992;
Holmes and Purdy 1992; Sherwin 1992; Wolf 1996c).

Thus the damage done by a bioethics that erases difference occurs
on a number of levels. Individual cases are wrongly construed, entire
patterns of profound harm are left unchallenged, bioethics itself becomes complicit in those harms, and the field devolves into a bioethics
by and for those who least need it—the already dominant. It is small
wonder that the field has remained alarmingly white and nondiverse.

Continued Marginalization

The historic avoidance of work on race, ethnicity, and gender in bioethics has now given way. Collections have begun to appear on African
American perspectives in bioethics (Dula and Goering 1994; Flack and
Pellegrino 1992), joining individual articles on the significance of race
(see, e.g., Conyers 1993; Dula 1994; Gamble 1993; Gamble and Blustein
1994; King 1992a and 1992b; Michel 1994; Randall 1993 and 1996). Empirical articles uncover correlations between ethnicity and attitudes
toward truth telling and clinical decision making (see, e.g., Blackhall et
al. 1995; Carrese and Rhodes 1995). Collections and monographs ap-

pear as well on gender and feminist perspectives in bioethics (see, e.g., Holmes and Purdy 1992; Mahowald 1993; Purdy 1996; Sherwin 1992; Tong 1997; Wolf 1996c). Bioethics journals publish gender-attentive pieces (see, e.g., Dresser 1992; Lebacqz 1991), anthologies include them (see, e.g., Cook 1994; Farley 1985), and symposia collect them (see, e.g., Colloquium 1995; Special Section 1996; Symposium 1996).

This greater attention to race, ethnicity, and gender is part of broader trends. Outside bioethics, medical journals document differences in health status, access to health care, and treatment by race, ethnicity, and gender (see, e.g., Ayanian et al. 1993; Hibbard and Pope 1986; Shaw et al. 1994; Svensson 1989; Wenger et al. 1993; Weisman 1987; Wenneker and Epstein 1989; Whittle et al. 1993). Mainstream organizations such as the American Medical Association report on differences in health care and access to experimental protocols by race and gender (Council on Ethical and Judicial Affairs 1990 and 1991). New legislation and regulation require greater inclusion of minorities and women in biomedical research (see, e.g., Centers for Disease Control 1995; 42 U.S. Code Annotated Supp. 1997). It would be hard for bioethics to ignore these developments.

But the greater attention to difference is also part of trends within bioethics. There is a shift away from the deductivism and acontextual universals of principlism (Wolf 1994b). Instead, the field is moving toward a greater inductivism, whether as an alternative or supplement to principlism. Bioethicists variously champion casuistry, narrative ethics, ethnography-driven bioethics, and other approaches (see, e.g., DeGrazia 1992; DuBose et al. 1994; Grodin 1995). It is all part of a greater attention to context and cases as the wellspring of bioethics analysis.

In keeping with this, there is an outpouring of empirical studies. We see research among the Navajo, for instance, on whether people really want to discuss death and engage in advance treatment planning (Carrese and Rhodes 1995). Other researchers compare Korean Americans, Mexican Americans, African Americans, and whites on whether people believe that patients should be told poor prognoses and make treatment decisions themselves in an exercise of individual autonomy (Blackhall et al. 1995). These and other studies directly challenge traditional assumptions in bioethics that certain principles are regnant and should apply to all.

With bioethics confronting this empirical outpouring and methodological ferment, one would think that race, ethnicity, and gender were finally assuming their proper place at the center of bioethics analysis. Unfortunately, that is not the case.

The core of bioethics remains largely untouched. Instead, the empirical studies seem merely to generate discussion of ethics for Navajos,

Mexican Americans, other ethnic groups, and sometimes women. A common approach is to suggest that patients can exercise their autonomy to choose the bioethical approach of their group (for example, family decision making instead of individual choice) (see, e.g., Gostin 1995). We bioethicists then do no further analysis, either of the dominant societal bioethics or that of the smaller group. The field thus becomes balkanized. There are satellite ethics discussed for various ethnic groups and women, but at the center, mainstream bioethics remains undisturbed.

The treatment of gender is an example. Gender-attentive work in bioethics is still often dismissed as a special conversation, a special ethics for women or feminists. Moreover, this ethics is frequently stereotyped as the ethics of care or as a revival of attention to the moral emotions,[2] even though much feminist work has been critical of the ethics of care and goes well beyond the moral emotions (see, e.g., Blum 1988; Sherwin 1992; Tronto 1993).

Further, gender-attentive work is often seen as writing about women, even though that work generally problematizes the relationship between the biological fact of sex and the social construction of gender. Feminists such as Catharine MacKinnon analyze gender as a system for subordinating certain people (MacKinnon 1987). Other authors write about this category as including not just women (Case 1995; Valdes 1995). People can be gendered female and coded as women, assigned attributes historically associated with the subordinated gender. Thus men can be gendered female: poor men, sick men, gay men, disabled men. But within bioethics, gender is rarely analyzed as a general category of privilege and social organization, with the capacity to shed a powerful light on the history of medicine and the biomedical sciences. Nor is the literature on the relationship of social gender and biological sex explored, despite the potential lessons for a field concerned with the relationship of physical or genetic makeup and social consequences.[3] The teachings of gender-attentive and feminist work are rarely embraced by mainstream bioethics.

The marginalization of such work as by the other, for the other, and about the other is unfortunately not confined to the domain of gender. Race-attentive work also has not received the attention it deserves in bioethics. Consider Dorothy Roberts's essay, "Reconstructing the Patient: Starting with Women of Color" (Roberts 1996). Her title derives from the fact that bioethics publications considering race generally do so as an afterthought. Thus they may analyze the situation of women and then note that women of color have a worse version of those problems, or they may present the situation of poor people dependent on government entitlement programs and then remark that poor people of color experience a more severe version. This assumes that the same harms and wrongs are inflicted on all regardless of race and that race is merely an aggravating factor. Roberts challenges that assumption. She argues that poor women of color receiving their health care in public settings experience special problems. Central bioethics tenets of confidentiality, truth telling, and respect for patient autonomy are not observed or even seriously advocated for such patients. Roberts substantiates this by examining the failure to insist on confidentiality when women are suspected of drug use during pregnancy, the abandonment of truth telling about the option of abortion in public clinics as approved by the Supreme Court in *Rust v. Sullivan*, and the failure to respect women's refusals of cesarean sections. Roberts argues that the core teachings of bioethics utterly ignore the experiences of poor women of color. She shows that bioethics proceeds from a certain narrow perspective. It is a bioethics for the privileged, the insured, and the white. It overlooks a powerful critique of the U.S. health care system to be found in the experiences of poor women of color (see also Scales-Trent 1991).

Roberts's criticism of bioethics is fundamental. She challenges the field's entire orientation and offers a new, more productive one grounded in the experiences of those the health care system has most marginalized. This is the sort of challenge that much work on race, ethnicity, and gender can offer. And bioethics should eagerly embrace this challenge. After all, the newer, more inductive bioethics now trumpeted as an improvement on the old principlism is supposed to be bottom *up*. That means the empirical data, thick case description, and rich ethnographies are supposed to lead somewhere—to a reformulation of governing norms, a revision of the casuistic paradigm, or a respecification of principles. Work attentive to race, ethnicity, and gender should be powering a reconceptualization of the bioethics core.

The fact that mainstream bioethics remains largely untouched by issues of race, ethnicity, and gender should tell us something. The field has clung to its liberal roots. It manages to recognize claims of difference when they are framed as a failure of equal treatment. If the problem is exclusion of minorities and women from research protocols, for instance, bioethics can see this as a violation of traditional concepts of justice. This requires no change in the usual bioethics principles and outlook, merely the application of a liberal insistence that like be treated alike.

But discussion of the meaning of difference has advanced much further than this in fields outside of bioethics (see Wolf 1995). Mary Joe Frug's writing, for example, divides work on gender into stages, with women first seeking to be treated the same as men, then seeking respect for differences, then transcending the sameness/difference debate by focusing on how power functions, and finally turning to contextualized and antiessentialist analysis of women's lives (Frug 1992, 3–11; see also Minow 1989, 2–4). Kimberlé Crenshaw offers a critique of the usual conceptualization of race and gender as two separate axes of difference, arguing that this systematically advantages the most privileged in each group, thereby "eras[ing] Black women" (Crenshaw 1989, 140). The literature is abundant and complex.

Yet bioethics remains largely isolated from such writings and concerns. There is little discussion within bioethics about alternative ways of looking at difference and how our field should approach it. We are failing to learn from the burgeoning literature on difference in law, philosophy, and politics, to name only three. Bioethics remains a field permeated by problems of difference yet deaf to the urgent and sophisticated dialogue on difference all around us.

The Future: Reinventing Bioethics

Roberts's work points to the future we should seek for bioethics: not the creation of satellite bioethics for specific groups or the simple denial of difference in the insistence that all be treated as whites or men, but serious engagement with difference and reintegration of bioethics at its core. There is some encouraging work in this vein. One example is Jorge Garcia's chapter in *African-American Perspectives on Biomedical Ethics* (Garcia 1992). He analyzes what it means to talk about an African American perspective, finds it a philosophically coherent notion, and suggests how that perspective would change the content of bioethics.

Janet Farrell Smith, writing in a feminist vein, similarly poses a fundamental challenge to the usual vision of how bioethics norms are generated (Farrell Smith 1996). Using Seyla Benhabib's revision of Jürgen Habermas's communicative ethics, Farrell Smith suggests that bioethics norms are legitimately created only through discussion involving the specific patient under conditions of communicative equality. This stands in stark contrast to the process for generating norms that has dominated bioethics thus far: expert specification by bioethicists themselves, in collaboration with medical and legal experts but rarely patients.

Roberts, Garcia, and Farrell Smith demonstrate the potential depth and range of bioethics work attentive to race, ethnicity, and gender. Such work will ask questions that go to the heart of bioethics: How did the moral scandals foundational to bioethics ever happen? How can human experimentation avoid exploitation of research subjects in the future? How can health professionals render good care in the face of profound differences between them and their patients? And how can health care plans and systems render care to diverse populations equitably and accountably?

Such work will also draw on broader literatures such as critical race studies and feminist jurisprudence that wrestle with questions plaguing bioethics as well (see, e.g., Delgado and Stefancic 1993; Weisberg 1993). Those literatures have grappled with the nature of rights, the reconceptualization of justice, the relationship of the biological and the cultural, and how one integrates cases or stories and norms. But those literatures examine these questions in a way that attends to difference, rather than simply erasing it. Bioethics has much to learn from those efforts.

A bioethics attentive to race, ethnicity, and gender must be careful; those differences have been used to harm people before and may be used to harm them again (King 1992a). But the new bioethics should not respond to the history of scandal and prejudice by simply ignoring group differences when they matter. Nor should the new bioethics stereotype people by seeing them automatically as members of various groups. Instead, it should explore how each individual constructs her identity, the meaning of group identification to the individual, the ways in which larger patterns of societal disadvantage figure in that individual's experience, and the harm done by those patterns themselves.

This change will not come easily or fast. Work on race, ethnicity, and gender was historically ignored in bioethics for deep structural reasons. That work is now held at arm's length for similarly profound reasons. And if such work succeeds in bioethics, it will not be by merely creating satellite ethics for different groups or by erasing difference. Instead, it will be by catalyzing core upheaval and change.

Notes

This chapter benefited from comments when earlier versions were presented at the World Congress of Bioethics, at the joint annual meeting of the Society for Health and Human Values and the Society for Bioethics Consultation, and at Michigan State University. Special thanks to Marion Secundy. Ryan Johnson of the University of Minnesota Law School provided research assistance.

1. In health care analyses, class has two components, socioeconomic status and insurance status, as those without health insurance in the United States tend not to be the poorest but the working poor who are not quite impoverished enough to qualify for entitlement programs.

2. The most recent edition of *Principles of Biomedical Ethics* (Beauchamp and Childress 1994) takes the salutary step of paying attention to feminist work for the first time, but it seems to fall into the trap I describe.

3. An exception is a chapter by Adrienne Asch and Gail Geller offering a feminist analysis of genetics (Asch and Geller 1996).

References

Andrews, Lori B. 1997. "Past as Prologue: Sobering Thoughts on Genetic Enthusiasm." *Seton Hall Law Review* 27: 893–918.

Annas, George J., and Michael Grodin, eds. 1992. *The Nazi Doctors and the Nuremberg Code: Human Rights in Human Experimentation.* New York: Oxford University Press.

Asch, Adrienne, and Gail Geller. 1996. "Feminism, Bioethics, and Genetics." In *Feminism & Bioethics: Beyond Reproduction,* ed. Susan M. Wolf. New York: Oxford University Press.

Ayanian, John Z., et al. 1993. "Racial Differences in the Use of Revascularization Procedures After Coronary Angiography." *Journal of the American Medical Association* 269: 2642–2646.

Banks, Taunya Lovell. 1989–1990. "Women and AIDS—Racism, Sexism, and Classism." *New York University Review of Law and Social Change* 17: 351–385.

Barker-Benfield, G. J. 1976. *The Horrors of the Half-Known Life: Male Attitudes Toward Women and Sexuality in Nineteenth-Century America.* New York: Harper Colophon Books.

Beauchamp, Tom L., and James F. Childress. 1994. *Principles of Biomedical Ethics.* 4th ed. New York: Oxford University Press.

Benson, Veronica, and Marie A. Marano. 1994. "Current Estimates from the National Health Interview Survey, 1993." *Vital and Health Statistics,* series 10 (190): 114–142.

Bhopal, Raj. 1997. "Is Research into Ethnicity and Health Racist, Unsound, or Important Science?" *British Medical Journal* 314: 1751–1756.

Blackhall, Leslie J., et al. 1995. "Ethnicity and Attitudes Toward Patient Autonomy." *Journal of the American Medical Association* 274: 820–25.

Blum, Lawrence A. 1988. "Gilligan and Kohlberg: Implications for Moral Theory." *Ethics* 98: 472–491.

Brandt, Allan M. 1985. "Racism and Research: The Case of the Tuskegee Syphilis Study." In *Sickness and Health in America,* ed. Judith Walzer Leavitt and Ronald L. Numbers. Madison: University of Wisconsin Press.

Caplan, Arthur L., ed. 1992. *When Medicine Went Mad: Bioethics and the Holocaust.* Totowa, NJ: Humana Press.

Carrese, Joseph A., and Lorna A. Rhodes. 1995. "Western Bioethics on the Navajo Reservation: Benefit or Harm?" *Journal of the American Medical Association* 274: 826–829.

Case, May Anne C. 1995. "Disaggregating Gender from Sex and Sexual Orientation: The Effeminate Man in the Law and Feminist Jurisprudence." *Yale Law Journal* 105: 1–105.

Centers for Disease Control. 1995. "Policy on the Inclusion of Women and Racial and Ethnic Minorities in Externally Awarded Research." *Federal Register* 60: 47, 947–951.

Colloquium. 1995. "Gender, Law and Health Care." *Maryland Law Review* 54: 473–632.

Conyers, John, Jr. 1993. "Principles of Health Care Reform: An African-American Perspective." *Journal of Health Care for the Poor and Underserved* 4: 242–249.

Cook, Rebecca J. 1994. "Feminism and the Four Principles." In *Principles of Health Care Ethics* ed. Raanan Gillon and Ann Lloyd. New York: Wiley.

Council on Ethical and Judicial Affairs. 1990. "Gender Disparities in Clinical Decision Making." *Journal of the American Medical Association* 266: 559–562.

———. 1991. "Black-White Disparities in Health Care." *Journal of the American Medical Association* 263: 2344–2346.

Crenshaw, Kimberlé. 1989. "Demarginalizing the Intersection of Race and Sex: A Black Feminist Critique of Antidiscrimination Doctrine, Feminist Theory, and Antiracist Politics." *University of Chicago Legal Forum*: 139–167.

Daniels, Norman. 1985. *Just Health Care*. New York: Cambridge University Press.

DeGrazia, David. 1992. "Moving Forward in Bioethical Theory: Theories, Cases, and Specified Principlism." *Journal of Medicine and Philosophy* 17: 511–539.

Delgado, Richard, and Jean Stefancic. 1993. "Critical Race Theory: An Annotated Bibliography." *Virginia Law Review* 79: 461–516.

Dresser, Rebecca. 1992. "Wanted: Single, White Male for Medical Research." *Hastings Center Report* 22 (January–February): 24–29.

DuBose, Edwin R., et al., eds. 1994. *A Matter of Principles? Ferment in U.S. Bioethics*. Valley Forge, PA: Trinity Press International.

Dula, Annette. 1994. "African American Suspicion of the Healthcare System Is Justified: What Do We Do About It?" *Cambridge Quarterly of Healthcare Ethics* 3: 347–357.

Dula, Annette, and Sara Goering, eds. 1994. *"It Just Ain't Fair": The Ethics of Health Care for African Americans*. Westport, CT: Praeger.

Ehrenreich, Barbara, and Dierdre English. 1973. *Complaints and Disorders: The Sexual Politics of Sickness*. Old Westbury, NY: Feminist Press.

———. 1979. *For Her Own Good: 150 Years of Experts' Advice to Women*. Garden City, NY: Anchor Books.

Farley, Margaret. 1985. "Feminist Theology and Bioethics." In *Theology and Bioethics: Exploring the Foundation and Frontiers*, ed. Earl E. Shelp. Boston: D. Reidel.

Farrell Smith, Janet. 1996. "Communicative Ethics in Medicine: The Physician-Patient Relationship." In *Feminism & Bioethics: Beyond Reproduction*, ed. Susan M. Wolf. New York: Oxford University Press.

Flack, Harley E., and Edmund D. Pellegrino, eds. 1992. *African-American Perspectives on Biomedical Ethics*. Washington, DC: Georgetown University Press.

Frug, Mary Joe. 1992. *Postmodern Legal Feminism*. New York: Routledge.

Gamble, Vanessa Northington. 1993. "A Legacy of Distrust: African Americans and Medical Research." *American Journal of Preventive Medicine* 9: 35–38.

Gamble, Vanessa Northington, and Bonnie Ellen Blustein. 1994. "Racial Differences in Medical Care: Implications for Research on Women." In *Women and Health Research: Ethical and Legal Issues of Including Women in Clinical Studies*, vol. 2, ed. Anna C. Mastroianni et al. Washington, DC: National Academy Press.

Garcia, Jorge L. A. 1992. "African-American Perspectives, Cultural Relativism, and Normative Issues: Some Conceptual Questions." In *African-American Perspectives on Biomedical Ethics*, ed. Harley E. Flack and Edmund D. Pellegrino. Washington, DC: Georgetown University Press.

Gostin, Lawrence O. 1995. "Informed Consent, Cultural Sensitivity, and Respect for Persons." *Journal of the American Medical Association* 274: 844–845.

Grillo, Trina, and Stephanie M. Wildman. 1991. "Obscuring the Importance of Race: The Implication of Making Comparisons Between Racism and Sexism (or Other-isms)." *Duke Law Journal*: 397–412.

Grodin, Michael A., ed. 1995. *Meta Medical Ethics: The Philosophical Foundations of Bioethics*. Boston, MA: Kluwer.

Guralnik, Jack M., and Suzanne G. Leveille. 1997. "Annotation: Race, Ethnicity, and Health Outcomes—Unraveling the Mediating Role of Socioeconomic Status." *American Journal of Public Health* 87: 728–730.

Haney López, Ian F. 1994. "The Social Construction of Race." *Harvard Civil Rights–Civil Liberties Law Review* 29: 1–62.

Hibbard, Judith H., and Clyde R. Pope. 1986. "Another Look at Sex Differences in the Use of Medical Care: Illness Orientation and the Types of Morbidity for Which Services Are Used." *Women and Health* 11 (Summer): 21–36.

Holmes, Helen Bequaert, and Laura M. Purdy, eds. 1992. *Feminist Perspectives in Medical Ethics*. Bloomington: Indiana University Press.

Horton, Jacqueline A., ed. 1992. *The Women's Health Data Book: A Profile of Women's Health in the United States*. Washington, DC: Elsevier.

Ikemoto, Lisa C. 1997. "The Racialization of Genomic Knowledge." *Seton Hall Law Review* 27: 937–950.

Jecker, Nancy S. 1993. "Can an Employer-Based Health Insurance System Be Just?" *Journal of Health Politics, Policy & Law* 18: 657–673.

Jones, James H. 1993. *Bad Blood: The Tuskegee Syphilis Experiment: A Tragedy of Race and Medicine*. 2d ed. New York: Free Press.

Kevles, Daniel J. 1992. "Out of Eugenics: The Historical Politics of the Human Genome." In *Code of Codes: Scientific and Social Issues in the Human Genome Project*, ed. Daniel J. Kevles and Leroy Hood. Cambridge, MA: Harvard University Press.

King, Patricia A. 1992a. "The Dangers of Difference." *Hastings Center Report* 22 (November–December): 35–38.

———. 1992b. "The Past as Prologue: Race, Class, and Gene Discrimination." In *Gene Mapping: Using Law and Ethics as Guides*, ed. George J. Annas and Sherman Elias. New York: Oxford University Press.

Lawrence, Charles R., III. 1987. "The Id, the Ego, and Equal Protection: Reckoning with Unconscious Racism." *Stanford Law Review* 39: 317–388.

Lebacqz, Karen. 1991. "Feminism and Bioethics: An Overview." *Second Opinion* 17 (October): 11–25.

MacKinnon, Catharine A. 1987. *Feminism Unmodified: Discourses on Life and Law.* Cambridge, MA: Harvard University Press.

Mahowald, Mary B. 1993. *Women and Children in Health Care: An Unequal Majority.* New York: Oxford University Press.

Merton, Vanessa. 1996. "Ethical Obstacles to the Participation of Women in Biomedical Research." In *Feminism & Bioethics: Beyond Reproduction*, ed. Susan M. Wolf. New York: Oxford University Press.

Michel, Vicki. 1994. "Factoring Ethnic and Racial Differences into Bioethics Decision Making." *Generations* (Winter): 23–26.

Minow, Martha. 1989. "Introduction: Finding Our Paradoxes, Affirming Our Beyond." *Harvard Civil Rights–Civil Liberties Law Review* 24: 1–7.

Nelson, Hilde Lindemann, and James Lindemann Nelson. 1996. "Justice in the Allocation of Health Care Resources: A Feminist Account." In *Feminism & Bioethics: Beyond Reproduction*, ed. Susan M. Wolf. New York: Oxford University Press.

Purdy, Laura M. 1996. *Reproducing Persons: Issues in Feminist Bioethics.* Ithaca, NY: Cornell University Press.

Randall, Vernellia R. 1993. "Racist Health Care: Reforming an Unjust System to Meet the Needs of African-Americans." *Health Matrix* 3: 127–194.

———. 1996. "Slavery, Segregation and Racism: Trusting the Health Care System Ain't Always Easy! An African American Perspective on Bioethics." *Saint Louis University Public Law Review* 15: 191–235.

Roberts, Dorothy E. 1996. "Reconstructing the Patient: Starting with Women of Color." In *Feminism & Bioethics: Beyond Reproduction*, ed. Susan M. Wolf. New York: Oxford University Press.

———. 1997. *Killing the Black Body: Race, Reproduction, and the Meaning of Liberty.* New York: Pantheon Books.

Savitt, Todd L. 1978. *Medicine and Slavery: The Diseases and Health Care of Blacks in Antebellum Virginia.* Chicago: University of Illinois Press.

———. 1982. "The Use of Blacks for Medical Experimentation and Demonstration in the Old South." *Journal of Southern History* 48: 331–48.

Scales-Trent, Judy. 1991. "Women of Color and Health: Issues of Gender, Community, and Power." *Stanford Law Review* 43: 1357–1368.

Shaw, Leslee J., et al. 1994. "Gender Differences in the Noninvasive Evaluation and Management of Patients with Suspected Coronary Artery Disease." *Annals of Internal Medicine* 120: 559–566.

Sherwin, Susan. 1992. *No Longer Patient: Feminist Ethics and Health Care.* Philadelphia: Temple University Press.

Special Section. 1996. "Feminist Approaches to Bioethics." *Journal of Clinical Ethics* 7 (1): 13–47.

Starr, Paul. 1982. *The Social Transformation of American Medicine.* New York: Basic Books.

Svensson, Craig K. 1996. "Representation of American Blacks in Clinical Trials of New Drugs." *Journal of the American Medical Association* 261: 263–265.

Symposium. 1996. "Feminist Perspectives on Bioethics." *Kennedy Institute of Ethics Journal* 6: vii–103.

Tong, Rosemarie. 1997. *Feminist Approaches to Bioethics: Theoretical Reflections and Practical Applications.* Boulder, CO: Westview Press.

Tronto, Joan C. 1993. *Moral Boundaries: A Political Argument for an Ethic of Care.* New York: Routledge.

42 U.S. Code Annotated 42 §289a–2 (Supp. 1997).

Valdes, Francisco. 1995. "Queers, Sissies, Dykes, and Tomboys: Deconstructing the Conflation of 'Sex,' 'Gender,' and 'Sexual Orientation' in Euro-American Law and Society." *California Law Review* 83: 1–377.

Weisberg, D. Kelly. 1993. *Feminist Legal Theory: Foundations.* Philadelphia: Temple University Press.

Weisman, Carol S. 1987. "Communication Between Women and Their Health Care Providers: Research Findings and Unanswered Questions." *Public Health Reports* 102 (Supp.): 147–151.

Wenger, Nanette K., et al. 1993. "Cardiovascular Health and Disease in Women." *New England Journal of Medicine* 329: 247–256.

Wenneker, Mark B., and Arnold M. Epstein. 1989. "Racial Inequalities in the Use of Procedures for Patients with Ischemic Heart Disease in Massachusetts." *Journal of the American Medical Association* 261: 253–257.

Whittle, Jeff, et al. 1993. "Racial Differences in the Use of Invasive Cardiovascular Procedures in the Department of Veteran Affairs Medical System." *New England Journal of Medicine* 329: 621–627.

Witzig, Ritchie. 1996. "The Medicalization of Race: Scientific Legitimation of a Flawed Social Construct." *Annals of Internal Medicine* 125: 675–679.

Wolf, Susan M. 1994a. "Health Care Reform and the Future of Physician Ethics." *Hastings Center Report* 24 (March–April): 28–41.

———. 1994b. "Shifting Paradigms in Bioethics and Health Law: The Rise of a New Pragmatism." *American Journal of Law & Medicine* 20: 396–415.

———. 1995. "Beyond 'Genetic Discrimination': Toward the Broader Harm of Geneticism." *Journal of Law, Medicine & Ethics* 23: 345–353.

———. 1996a. "Bioethics: From Mirror to Window." *Saint Louis University Public Law Review* 15: 183–189.

———. 1996b. "Introduction: Gender and Feminism in Bioethics." In *Feminism & Bioethics: Beyond Reproduction*, ed. Susan M. Wolf. New York: Oxford University Press.

Wolf, Susan M., ed. 1996c. *Feminism & Bioethics: Beyond Reproduction.* New York: Oxford University Press.

Part II
Reproduction and Beyond

5

Abortion, Chernobyl, and Unanswered Genetic Questions

Laura Shanner

Exposure to radiation has long been known to cause genetic damage, and exposure to the fallout of a nuclear reactor meltdown would understandably raise concerns about genetic anomalies in offspring conceived after the exposure. Such fears have persisted in Eastern Europe following the explosion in Unit 4 of the Chernobyl nuclear power plant in Ukraine[1] (then USSR) on April 26, 1986. On the tenth anniversary of the disaster, an international convention to review the available information on the consequences of the Chernobyl accident almost completely ignored the topic of radiation-induced genetic anomalies but concluded nevertheless that genetic fears were overblown and unfounded. We should not be satisfied with these conclusions. The little that was said at the conference about genetic effects is logically flawed and methodologically suspect. In addition, several studies existing at the time indicate evidence of radiation-induced genetic anomalies, but these studies and their authors were not included in the conference agenda.

It is a philosopher's work not to document scientific data but to identify fallacious arguments, hidden or suppressed assumptions, and unsupported conclusions. Sadly, the tenth-anniversary conference provided numerous examples of how one ought not to construct an argument to reassure concerned victims of radiation exposure. Of particular concern is the failure to consider evidence regarding rates of genetic anomalies in the context of reproductive choices made in the region. As a feminist, I note that these oversights will have particularly adverse effects for women, because genetic worries increase the likelihood of invasive prenatal genetic testing and abortion. I also note with dismay that legitimate questions about health effects for women and future children were given far less attention than the financial and political implications of the reactor meltdown. I believe that the full story of Chernobyl's effects has not been told, and unless research priorities are changed soon, it may never be fully known. We can conclude, however, that the assurances offered at the tenth-anniversary conference are inadequate at best and are very likely misleading. An international review of the existing data and greater efforts to document the effects are required to ease the minds of the residents, and especially the women, of the region.

Map Showing the Study Area

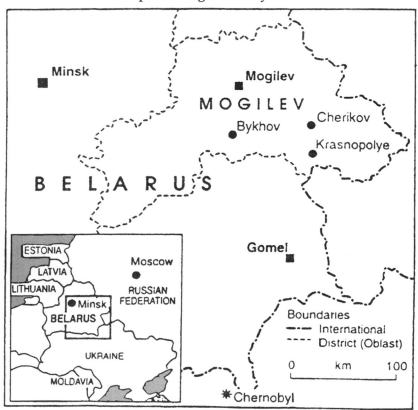

Source: Y. E. Dubrova et al. 1996. "Human Minisatellite Mutation Data After the Chernobyl Accident." *Nature* 380 (25 April).

Context of the Problem: Daily Life in the Former USSR

The mid-1980s to the mid-1990s has been a decade of extraordinary upheaval and uncertainty for the people of the former Soviet Union, as their governments and economies have undergone massive restructuring from totalitarian communism to democratic elections and open markets. These transitions have been difficult and certainly are not complete. Russia and other states of the former Soviet Union continue to struggle with economic and political reforms, poverty, collapsing social programs, inconsistent supply lines, and political instability.

The cradle-to-grave public support system under the Communist government has evaporated; taken-for-granted securities for citizens in a socialist state are now frequently lacking or exist in new and confusing forms. For example, one colleague in Moscow described feeling overwhelmed by the need to begin saving for his upcoming retirement after a lifetime of assurances that the government pension would fully support him in his old age. Those of us in the West who are routinely perplexed by international currency conversions, income tax laws, and the vagaries of the stock market can only imagine the distress of a late-career realization that our survival now depends upon immediate mastery of investment strategies that we never even knew existed.

This uncertainty and upheaval affects all aspects of the average citizen's life through health care restructuring, employment and market redesign, the loss of subsidized housing, unstable currency, and rampant inflation. Workers still receive very low wages compared to Western standards, and even these minimal wages are frequently in arrears by weeks or months due to cash flow shortages and economic instability. Long queues at Soviet stores with empty shelves are largely gone now, as many items are available and international chains are opening franchises in major cities; unfortunately, many people cannot afford the prices of these imported items.

Reproductive Patterns

Amid such great political, economic, and social uncertainty, it is not surprising that the birthrates in the countries of the former USSR are suppressed; indeed, several regions report far more deaths than births, a demographic indicator commonly (but not always) associated with serious social instability. When one's own economic survival is uncertain, it is understandable to think it unwise to bring a child—another mouth to feed—into the household; if there is pessimism about the future, it may seem unreasonable to introduce a loved and wanted child to a hard and chaotic life.

In October 1994, I visited the Ott Institute of Obstetrics and Gynecology in St. Petersburg, considered the premier obstetrics/gynecology facility in all of Russia and at one time among the best in all of Europe. I was told that the delivery rate had fallen by over 60 percent in recent years. For decades, women traveled great distances to give birth at the Ott Institute, which could handle approximately six thousand deliveries per year and had to turn hundreds of women away for lack of space. By the mid-1990s, however, the birthrate had fallen to approximately two thousand deliveries per year. If the best and best-known obstetric facility in the country suffers such a drastic reduction in birthrates, the clinicians suggested, then other facilities must be facing similar or even greater reductions. Sharp declines in birthrates have been observed in the areas most affected by the Chernobyl disaster as well, as part of a larger pattern of demographic distress that includes the migration of workers from contaminated to uncontaminated areas.

The Ott Institute also documented a dramatic rise in the number of genetic tests performed during pregnancy and noted that even with nondirective counseling, patients almost always aborted a pregnancy in which an anomaly was detected. Statistics from the Ott prenatal testing unit reveal the following numbers of tests done per year:

Year	Total No. of Tests (Amniocentesis, Chorion Biopsy, and Chorionic Villus Biopsy)
1987	98
1988	88
1989	126
1990	255
1991	275
1992	350
1993	352
Jan.–Apr. 1994	109

Interpreting this data is difficult. Part of the explanation for the increase in genetic testing, as well as for the low birthrate, may reflect increased fear of genetic abnormality as a result of Chernobyl. A series of 1991 studies (Odlind and Ericson 1991; Knudson 1991; Spinelli and Osborn 1991; Irgens et al. 1991) document increases in the abortion rates in Sweden, Denmark, and Italy in 1986–1987 but no increase in abortions in Norway for the same time period. The slight reduction in tests performed at the Ott Institute between 1987 and 1988 may reflect a temporary increase in 1987 of worries about genetic damage from Chernobyl that were relieved somewhat after another year had passed. It is also possible that an increase in genetic test requests and abortions in Russia expected in 1986–1987 may have been delayed by a lack of public awareness of exposure due to the government's cover-up of the explosion at the time it occurred.

We should note, however, the rapid quadrupling of tests between 1988 and 1992 and the stable plateau at a high level for 1992–1994. This dramatic increase coincides less with the fallout of Chernobyl than with the period of greatest upheaval in the Soviet states. Staff at the Ott Institute considered economic insecurity a primary motive for genetic testing and abortion, at least in St. Petersburg. The low birthrate indicates that few people are willing to expand their families amid the current societal upheaval; without secure medical and social supports, raising a child with genetic or chromosomal abnormalities would be too great a burden even for those with initially high optimism and resources. Fears of genetic risk from Chernobyl and other sources therefore must be considered in light of, and are likely worsened by, larger patterns of social and economic insecurity.

Unfortunately, the political and economic instability that appears to explain the drop in the birthrate and perhaps also the rise in the rate of genetic testing has made it difficult for women in Eastern Europe to obtain effective contraceptives to prevent pregnancy. Imported condoms are often expensive and difficult to find, while locally produced ones are of unreliable quality. There is also, among some segments of the population, a feeling of revelry in new-found freedoms that makes condoms unattractive to use; the result is a skyrocketing HIV infection rate, as well as an increased risk of pregnancy. Hormonal contraceptives may not be consistently available, and apparently, identical refills may actually be slightly different doses or formulations from different suppliers or nations, thus increasing the likelihood of accidental pregnancy. Contraceptives of any form, not to mention basic feminine hygiene products that are taken for granted in Western countries, are therefore frequently considered luxury items. Not surprisingly, contraceptive use is often inconsistent, and contraceptive failure rates are high.

The health impact for women of the unstable economic outlook, the inconsistent availability of contraceptives, changing social and sexual norms, and the increased fear of genetic risk is tremendous. Clinicians at the Ott Institute commented that although confirmed statistics are hard to find, they estimated that a typical Russian woman may have five to seven abortions during her reproductive lifetime.

The Tenth-Anniversary Conference on Chernobyl's Effects

On April 8–12, 1996, almost exactly ten years after the Chernobyl disaster, the European Community (EC), the World Health Organization (WHO) and the International Atomic Energy Agency (IAEA) held a conference in Vienna, Austria, titled "One Decade After Chernobyl: Summing Up the Consequences of the Accident." The purpose of the conference was to summarize available information about the consequences of the accident, to provide guidance for preventing such disasters in the future, and to learn how to prepare better responses to nuclear disasters if or when they do occur.

A similar conference in 1991, also held in Vienna and sponsored by the IAEA, drew criticism from the Belarussian and Ukrainian delegations for "excessive optimism" regarding the official findings of limited health effects from Chernobyl (Webb 1991; Rich 1992). It was hoped that after another five years, better data would be accumulated, more studies would have been conducted into the short- and long-term health effects of radiation exposure, and the work done by local researchers would be incorporated into the data collection of Western and international research teams. To reach settled conclusions about health risks, either clear data should document heightened rates of increase of genetic abnormalities and/or other health complications, or data should convincingly show that there is no such increase. Unfortunately, however, the 1996 international conference again offered a highly optimistic picture of outcomes while failing to include existing data or provide support for the reassuring conclusions. The distress of making reproductive choices amid this uncertainty, as well as the health effects for women of repeated genetic testing and abortion, clearly have not been taken seriously enough at top international levels.

The first indication that genetic risks would not be adequately addressed in this conference comes from the listing of technical symposia in the *Summary of the Conference Results* (1996):

1. Clinically Observed Effects
2. Thyroid Effects
3. Longer-Term Health Effects
4. Other Health-Related Effects
5. Consequences for the Environment
6. Social, Economic, Institutional, and Political Impact
7. Nuclear Safety Remedial Measures
8. The Consequences in Perspective

One might expect that a topic as widely known and as emotionally charged as genetic abnormality in offspring following radiation exposure would receive its own focused seminar. Would people interested in finding data on genetic anomalies look to "clinically observed effects," "longer-term health effects," or "other health-related effects"? It turns out that hereditary disorders fit into the "longer-term health effects" category, although risks of genetic damage were also briefly addressed in the symposium on the environment.

"Clinically observed effects" focused solely on the effects of radiation for 600,000–800,000 occupationally exposed workers, including approximately 200,000 emergency workers or "liquidators" who were called in to clean up the disaster site in 1986–1987. The primary results here concern acute radiation syndrome (ARS): 237 exposed individuals were admitted to hospitals suffering from clinical symptoms attributed to radiation exposure. ARS was diagnosed in 134 cases. Of these 134 patients, 28 died of radiation-induced injuries within the first three months, and 3 others died of causes thought unrelated to radiation. Over the past ten years after the acute phase, an additional 14 have died (although the deaths do not appear to correlate with the original severity of ARS), and the rest are generally in poor health (*Summary* 1996, paras. 3, 12).

The other most noticeable health effect of Chernobyl, receiving its own technical symposium, is thyroid cancer in children born before or within six months of the accident. Approximately 800 cases had been diagnosed at the time of the conference, with over 400 of these in Belarus and 360 in Ukraine. Data from Belarus indicates a national rate of occurrence of 14.6 cases per million children; in the heavily contaminated Gomel region, the rate is 100 cases per million, which is 200 times the rate in uncontaminated Great Britain (*Technical Symposium* 1996, sess. 2).

In the "longer-term health effects" symposium, concerns about other radiation-induced health effects were largely discredited: "Apart from the dramatic increase in thyroid cancer in those exposed as children, there is no evidence to date of a major public health impact as a result of radiation exposure due to the Chernobyl accident in the three most affected countries (Belarus, Russia and Ukraine)" (ibid., sess. 4). Although there have been some reports of increases in incidence of specific non-thyroid malignancies in some contaminated areas, the reports are not consistent (*Summary* 1996, para. 24). The *Summary of the Conference Results* also notes several critical limitations to the data:

> The total expected excess fatalities due to leukaemia would be of the order of 470 among the 7.1 million residents of "contaminated" territories

and "strict control zones" which would be impossible to distinguish from the spontaneous incidence of about 25,000 fatalities. . . . In summary, to date, no consistent attributable increase had been detected either in the rate of leukaemia or in the incidence of any malignancies other than thyroid carcinomas. . . . Future increases over the natural incidence of all cancers, except for thyroid cancer, or hereditary effects among the public would be difficult to discern, even with large and well-designed long term epidemiological studies. . . . Increases in the frequency of a number of non-specific detrimental health effects other than cancer among exposed populations, and particularly among liquidators, have been reported. It is difficult to interpret these findings because exposed populations undergo a much more intensive and active follow-up of their state of health than does the general population. Any such increases, if real, might also reflect effects of stress and anxiety. (1996, paras. 25–27)

Appropriate and, indeed, impressive attention was paid to the psychological effects of the accident and to the context of social instability that increases psychosocial distress:

Past experience of accidents unrelated to radiation has shown that the psychological impact may persist for a long period. In fact, ten years after the Chernobyl accident, the evolution of symptoms has not ended. It can be expected that the importance of this effect will decrease with time. However, the continuing debate over radiation risks and countermeasures, combined with the fact that effects of the early exposures are now being seen (i.e. the significant rise in thyroid cancers among children), may prolong the symptoms. In evaluating the psychological impact, account should be taken of the psychological effects of the breakup of the USSR, and any forecast should take into account the economic, political, and sociological circumstances of the three countries. The symptoms such as anxiety associated with mental stress may be among the major legacies of the accident. (Ibid., para. 66)

On the other hand, this sensitivity to sociopolitical context and the psychological anxiety produced by the accident also provided grounds to dismiss several health concerns as hypochondriacal, psychosomatically induced, or simply unrelated to radiation exposures. It is certainly true that the causal effects of disease in a sociopolitically unstable region are extremely difficult to determine, and thus the warnings about methodological inaccuracy are well taken. However, it is also true that *one cannot conclude that there is no increased risk* based on the same complex, methodologically inadequate data. Accordingly, statements that seek to assure us of minimal risk fail in that attempt, because there are no clear grounds for reaching reassuring conclusions.

At best, one could conclude from complex sociopolitical context and multifactorial influences that we frankly do not know what sorts of risks are present in most conditions beyond the blatantly obvious ones of radiation illness and thyroid cancer. The statement that unofficially summarizes the technical symposium on "other health-related effects," however, dismisses possible health effects as follows (quoted in its entirety):

> Widespread public anxiety and pessimism about the Chernobyl accident among hundreds of thousands of people in the affected areas appears to be out of all proportion to the verifiable radiation induced health effects. This stress is nonetheless very real and has caused widespread damage to the general health and wellbeing of the population.
>
> A large proportion of the relevant inhabitants—whether evacuated or not—complain of ailments they believe to be due to radiation exposure. The level of general health is in any case low, and radiation fears are compounded by poor public understanding of radiation; initial secrecy; subsequent lack of effective communication; and the collapse of the former centralized political and economic systems. Distrust of "authorities" is widespread. It may be in this comparatively low-cost field of better communication that more needs to be done to help offset what is probably the most pervasive after-effect of the 1986 accident. (*Technical Symposium*, 1996, sess. 4)

The need for better communication and restoration of trust is quite clear. The limited and dismissive conclusions from the international conference, however, fail to communicate well or engender trust in the authorities charged with the task of compiling the relevant data. The official *Summary* document reinforces the obfuscation, rather than clarification, of psychological and sociopolitical factors in disease patterns:

> Different factors, such as economic hardship, are having a marked effect on the health of the population in general, including the various groups exposed as a result of the accident. The statistics for the exposed populations are being examined in the light of the clear general increase in morbidity and mortality in the countries of the former Soviet Union so as to preclude the misinterpretation of these trends as being due to the accident.
>
> The public perception of the present and future impact of the accident may have been exacerbated by the difficult socioeconomic circumstances in the USSR at the time, by the countermeasures that the authorities took to minimize the accident's impact, and by the public's impression of the risks from the continuing levels of radioactive contamination. (*Summary* 1996, paras. 64–65)

Of particular concern for the present discussion is the rate of hereditary disorders, which is given astoundingly short shrift. There is *no
mention at all* of genetic abnormalities in the official *Summary* document
or in the unofficial summaries of the technical symposia. Only a brief
statement about birth defects is made in the summary of a press
briefing: "Reports of birth defects should be regarded with care since
there is no reliable evidence of any significant change in numbers since
the accident." Genetics were supposed to be discussed in the third
technical symposium on "longer-term health effects," since "clinically
observed effects" focused on acute radiation syndrome, "thyroid effects" focused on thyroid cancer, and the "other health-related effects"
focused on psychological sequelae. What, then, was included in the
summary of the technical symposium that was supposed to investigate
genetic risks? The entire unofficial summary of the Third Topical Session reads as follows:

> Besides the dramatic [rise in] thyroid cancer in persons exposed when
> they were children, no evidence of a major public health impact is in evi-
> dence as a result of the radiation exposure in the Chernobyl accident in
> the three affected countries (Belarus, Russia or the Ukraine). It is believed
> that the major radiological impact of the accident will be greatest in the
> group known as the "liquidators" (those who worked at the Chernobyl
> plant and helped with the clean-up activities following the accident).
>   The conclusions reached in the session are that continued studies are
> needed in the future and they include:
>
> 1. Passive monitoring of the radiation exposed populations (in the
>   form of a population registry) from the Chernobyl accident to
>   prove data about disease patterns.
> 2. Focused studies of selected populations with exposures in the
>   low to medium radiation dose range.
> 3. Studies of radiation induced cancer risks through studies of the
>   thyroid cancers occurring in the young people. Although the
>   studies may not benefit today's patients from the Chernobyl acci-
>   dent, they will be valuable for protection of future populations
>   with possible radiation exposure. (*Technical Symposium* 1996, sess.
>   3)

Once again, there is no mention of genetic effects. It appears that several technical symposia each ignored genetic risk on an assumption
that someone else would address it, or perhaps that it was not worth
addressing. Despite the existence of separate technical symposia devoted to thyroid effects and ARS, these two obvious and undisputed
manifestations of radiation exposure dominated *all* of the panels that

were to investigate the entire range of human health effects of Chernobyl's radiation.

Existing Data

While needed data on genetic anomalies is lacking, several studies that were available prior to the conference were apparently disregarded. I must emphasize that a few worrisome studies are not conclusive evidence of a causal effect. However, the international conference needed to consider in much greater depth the available data, call for continuing research to establish causal relationships and settle ongoing uncertainties, and offer *convincing* data to allay those worries that are truly unfounded. Choosing to ignore the topic of genetic damage while merely dismissing the widespread public anxiety over risks is not an argument to support a conclusion that the risk is minimal.

Indeed, several studies demand our attention. One of the most compelling indicates a statistically significant, twofold increase in mutation frequency observed in offspring born between February and September 1994 to parents who have been exposed to radiation. Blood samples were taken from 79 families in the highly contaminated Mogilev region of Belarus and also from 105 families in uncontaminated regions of Britain. The parents in the study had resided in their respective regions since the time of the accident, and paternity/maternity were ensured before families were included in the study. DNA samples from the children were compared with samples from their parents; "mutant DNA" was defined as a fragment of the child's DNA unattributable to either parent and thus indicating spontaneous mutation. Mutant DNA was found twice as frequently in the Mogilev families as in the British families. This is believed to be the first clear experimental evidence that ionizing radiation can induce germ-line mutations in offspring, although the effects of such mutations are unknown because the stretches of DNA in which the mutations occurred have no known role in human health (Dubrova et al. 1996).

Additional studies provide other disturbing pieces of an unsolved puzzle. A statistically significant increase in trisomy 21 was observed in Germany in January 1987, exactly nine months after the Chernobyl accident (Sperling et al. 1991). The highest rates were observed in the more heavily contaminated, southern parts of Germany, although a supraregional study across all of Germany showed no significant effect (see methodological difficulty #4, below). A significantly higher incidence of Down's syndrome was observed in the Lothian region of Scotland in 1987, which is temporally associated with Chernobyl, but a biologically plausible explanation linking this phenomenon with Chernobyl remains lacking due to the low rate of radiation in Lothian (Ramsay, Ellis, and Zealley, 1991). In Norway, a positive association was observed between total radiation dose and hydrocephaly, while a negative association was found for Down's syndrome and no associations were found for small head circumference, congenital cataracts, anencephaly, spina bifida, or low birth weight (Lie 1992). A World Health Organization study released in 1995 indicates that children irradiated while in the womb suffer a greatly increased risk of mental retardation, behavioral disorders, and emotional problems (Edwards 1991). Genetic damage has also been clearly identified in voles, a rodent species common and still thriving in the Chernobyl region. Nine voles taken from inside the "hot zone" showed forty-six mutations in a single gene in mitochondrial DNA, while ten animals from outside the hot zone showed four mutations (Hillis 1996; Dickman 1995).

Interpreting the Convention's Omissions

The convention was correct in noting an abundance of studies showing conflicting results or no correlation between Chernobyl and genetic anomalies. However, when confronted with inconsistent data that includes some well-documented and deeply worrisome results, it would be prudent and more logically defensible to admit continuing uncertainty than to conclude that the worries are unfounded and overblown. As happened in 1991, existing data appears not to have been considered seriously by the 1996 international conference before being dismissed. It is therefore imperative that we consider the reasons given, and some possible reasons not contained in the conference documentation, for underplaying concerns about genetic risk.

Methodological Difficulties

As the conference concluded in discussions of nonthyroid cancer and other health effects, epidemiological data on health effects is quite difficult to interpret conclusively, especially in a context of poverty, political instability, and widespread anxiety. It is correct, therefore, to be humble and limited in one's claims about cause–effect relationships. Among the many problems confounding data collection, we need to consider the following:

1. Many regions affected by the blast lack accurate data on occurrences of particular conditions prior to the accident; current cases therefore cannot be seen to represent an increase or decrease, as there

is no baseline against which to compare them. Comparisons to baseline rates of disease in other regions may be misleading because the baseline rates in the test area and control area may have differed at the outset due to local conditions not related to radiation. As noted in the conference *Summary*, the high degree of postexposure monitoring of affected populations may also cause otherwise unnoticed cases to be diagnosed; there may be no actual increase from the baseline but simply a more accurate measuring of the baseline rate itself.

2. Given the large populations affected by the drifting cloud of radioactive particles and the relatively rare occurrence of many conditions, an increase in cases may still affect relatively small numbers of people and therefore may not be detectable. This point was made repeatedly in the conference summaries regarding nonthyroid cancers.

If a small increase is detected and reliably documented, then we face a further problem of interpreting it. Suppose that the incidence of a condition increases from 0.1 percent of the population to 0.2 percent; either this is a negligible 0.1 percent increase across the population that should reassure us that risks remain very low, or it represents a doubling in frequency that should prompt further investigation into the causal mechanisms of the condition. Further, even a tiny increase in the rate of a disease can have a devastating cumulative effect on health care resources and human suffering when this tiny statistical rise occurs over a population of hundreds of millions of people and causes thousands of cases.

3. Increases in common conditions are less easily noticed than increases in conditions perceived to be rare. It is also more difficult to isolate the proximate cause of particular cases when the condition is common and therefore caused by other (common) factors.

4. Data collected over large regions or populations may fail to distinguish the variety of dose exposures; clear correlations between high doses of radiation exposure and certain effects may therefore easily be lost. Recall, for example, the data on thyroid cancer cases: the highly contaminated Gomel region of Belarus showed a rate of 100 cases per million children, which is 200 times the baseline rate observed in Great Britain; Belarus as a whole shows only 14.6 cases per million, or a 35-fold increase over Britain; the total of 809 cases (424 of which occurred in Belarus) in all three affected nations is a higher rate than normal, but the correlation is not as frighteningly obvious as the data from Gomel.

5. There is ample evidence that many important diseases, including most cancers and cardiovascular disease, have multifactorial origins. Since this is true, a demonstrated rise in the rates of a particular condition may be merely coincident with the Chernobyl accident rather than

causally related. This complication was repeatedly raised in the *Summary* document, often in reference to stress-induced illness.

Although it is well known that stress can induce infertility and miscarriage, it is unclear whether stress and anxiety can promote genetic anomalies in offspring. Multifactorial influences such as stress and anxiety are less persuasive explanations for genetic risks than for cancer or many other health conditions.

6. Standards for research methodologies vary between Eastern and Western scientists, with the result that Eastern European research is often discounted in Western journals and conferences.

Political Influences

In addition to these methodological problems of data collection and interpretation, political pressures clearly have affected the collection and interpretation of data on risks of radiation exposure. The international conference declined to include these factors in their summaries:

7. Local data collection after the accident has been limited due to the extreme economic hardship and political upheaval in the affected nations. There is insufficient funding from local governments to conduct proper studies, and international donations have not been provided in sufficient amounts to conduct the research that is required. In my 1994 visit to Belarus, our delegation was repeatedly asked by members of the National Academy of Sciences, the Institute for Hereditary Diseases, and the School of Medicine to send donations of subscriptions to basic journals, funding for conference travel, computer equipment, and other very basic research resources. Meetings with top officials at these institutions frequently took place in cold, darkened rooms; the institutions could not afford to use electricity, and the central steam heating system in Minsk would not be turned on by the government until the temperature dropped below 40°F. Under these conditions of personal and professional deprivation, it is not surprising that researchers have been unable to collect the extensive data needed on genetic and other health conditions.

Although some researchers characterize Chernobyl research as a growth industry (Williams and Balther 1996), the research brought to the attention of the international conference seems limited to a few key topics. The failure of international governments and research institutions to support the needed data collection in this region is particularly disappointing after the 1991 international conference concluded that much more data needed to be collected.

8. There is some distrust among scientists of different cultures, with particular fears from former Soviet researchers that their hard work

will be stolen by outside researchers; as a result, much locally collected data is unavailable to outside researchers ("Truth and Chernobyl" 1991).

9. Political tensions among the former Soviet states may prevent cooperation in the region, and political tensions between former Soviet states and other nations may prevent the open dissemination of information with potential defense applications (ibid.).

10. We must consider the possibility that nations with nuclear capabilities and histories of nuclear experimentation (notably the United States, Russia, and France) may prefer not to delve too deeply into certain outcomes of radiation exposure. If extensive health effects are documented, these governments are more likely to be held liable for injuries caused by nuclear bomb detonations and, in the United States, research into radiation exposure on unsuspecting citizens. Blatantly obvious results such as ARS and thyroid cancer are hard to dismiss; most other, less obvious effects are easier to ignore. While I do not generally advocate conspiracy theories, we nevertheless ought to consider political and economic motivations not to uncover certain truths.

The Context of Pregnancy Decisions

Perhaps the most important mistake in leaping from consideration of very limited data on genetic anomalies to optimistic conclusions about genetic risk is the lack of attention paid to the social context of reproductive choices. As the *Summary* document rightly indicates, the most prevalent effects of Chernobyl include psychological distress; the region in which the accident occurred additionally suffers extensive sociopolitical upheaval, economic insecurity, and distrust of authorities. These factors must certainly be taken into account when evaluating the rate of genetic abnormalities in the population, but doing so leads to very different conclusions than those reached by the international conference.

The lack of any observed increase in the number of birth defects must be situated in the context of the region's low birthrate, high abortion rate, and high rates of genetic testing. Since very few babies are born at all due to political and economic uncertainty, and pregnancies that do go to term are increasingly likely to be screened for genetic or chromosomal abnormalities, then it follows that exceedingly few babies with congenital abnormalities will be born. As noted above, caring for an ill or disabled child is simply beyond the capacity of most citizens of the region who struggle to earn a living and who cannot count on state-provided health care and welfare support. Indeed, given the pressures on women in the former Soviet states to undergo genetic testing and to discontinue pregnancies, it is likely that there is a *decrease* in the recorded number of births with anomalies.

The inconsistent and incomplete data currently available on the occurrence of birth defects, by itself, fails to settle questions of the genetic risks involved. The inability to draw clear conclusions from an incomplete set of puzzle pieces frustrates studies on the frequency of most health conditions, but abortion further confounds data on reproductive outcomes. Data collected on birth anomalies *does not document the frequency of genetic abnormalities relative to conceptions*; rather, it documents reported cases of *births* with certain conditions, while aborted pregnancies or miscarried fetuses with those traits are not counted. Similarly, it is difficult to document trends in miscarriage rates when the rate of elective pregnancy termination is extremely high; many women may terminate a pregnancy prior to its eventual fated miscarriage, but we can never know what would have happened if the abortion had not been performed.

Conclusions

The international conference clearly failed to meet its mandate, which was to examine and summarize all of the known effects of Chernobyl's radiation on human populations and the environment. The failure to address genetic risks seriously is particularly frightening in light of the policy impact that this conference will have in regulation of nuclear facilities, health care interventions in the aftermath of another accidental exposure, and continuing research into the full extent of the health effects of radiation exposure. The preamble to the *Summary of the Conference Results*—the document that failed to mention *even once* any data on genetic effects beyond a passing observation that data will be difficult to collect—establishes in its second paragraph the enormous impact that this conference is intended to have: "The Joint Secretariat of the Conference [EC, IAEA, and WHO] recommends that this summary of the Conference results be used as the basis for decisions concerning future work and collaboration with the aim of alleviating the consequences of the Chernobyl accident." We may never know the true story about the effects of Chernobyl for offspring. Given the confusing, contradictory, and incomplete data, the complex social context in which reproductive decisions are made, and the multitude of stresses affecting residents of the former Soviet states, it seems unlikely that the causal mechanisms will be fully sorted. Perhaps the risks are not heightened at all, as the conference concluded, and fears of genetic risks are truly misplaced. It seems to me more plausible, however, that there are risks genuinely worthy of concern until they are convincingly laid to rest with accurate and comprehensive data that tracks genetic anomalies in miscarried and aborted fetuses as well as in births. Such studies have not, to my knowledge, been undertaken in the affected regions, and the international conference conclusions offer no support for such a thorough investigation into reproductive health.

This oversight poses a genuine threat to the health of offspring who may suffer undetected genetic damage and to women who undergo extensive prenatal testing and frequent abortions out of fear and the inability to cope with a disabled child. Even more frightening is the prospect that the conclusions from the tenth-anniversary conference on Chernobyl will be used to undermine concerns about genetic risk in future cases of radiation exposure, possibly leading to an increase in preventable birth anomalies in several populations around the world in future years. If the international conference conclusions are overly optimistic or inaccurate, then the damage caused by this precedent has the potential to be enormous. We cannot close the chapter on Chernobyl without addressing such serious consequences.

Note

1. Conventions for naming the countries require a brief note. I would argue that, as a general rule, we should show respect for nations and people by using the terms that they choose for themselves, rather than renaming them for our convenience. It is common in English to call Ukraine "the Ukraine," but this is inaccurate. Ukraine is an independent nation, not a region, adjective, or plural noun, and thus it takes no article; the English phrase "the Ukraine" is more like "the Canada" (incorrect) than "the Nile valley" or "the Cayman Islands." Using the article connotes "the Ukraine region of the USSR," which is inaccurate politically and may be perceived as offensive.

Belarus has had several spellings, some of which appear unchanged in quotes or citations for this article. I have chosen to adopt the spelling used on postage stamps and official documents issuing from the country itself. Much of the confusion is due to the translation between Cyrillic and Roman alphabets, and further evolution of the name has occurred with historical and political changes. Originally, Byelorussia or "white Russia" was a region of the Russian Empire and the USSR. With independence, the traditional name has evolved (in English) to Belarussia, Byelorus, Bielarus, and other similar variants, most of which reduce the linguistic connection to Russia.

References

Cardis, R., L. Anspach, V. K. Ivanov, I. Likthariev, K. Mabuchi, A. E. Okeanov, and A. Prisyazhniuk. 1996. Estimated long term health effects of the Chernobyl accident. Joint EC/IAEA/WHO International Conference: "One Decade After Chernobyl." *Technical Syposium, Unofficial Summaries,* Topical Session 3.

Dickman, Steven. 1995. Chernobyl's voles spring a genetic surprise. *New Scientist,* August 12, 14.

Dubrova,Y. E., V. N. Nesterov, N. G. Krouchinsky, V. A. Ostapenko, R. Neumann, D. L. Neil, A. J. Jeffreys. 1996. Human minisatellite mutation rate after the Chernobyl accident. *Nature* 380 (April 25): 683–686.

Edwards, Rob. 1991. Will it get any worse? *New Scientist* (December 9): 14–15.

Hillis, David M. 1996. Life in the hot zone around Chernobyl. *Nature* 380 (April 25): 665–666.

Irgens, L. M., R. T. Lie, M. Ulstein, T. Skeie Jensen, R. Skjaerven, F. Sivertsen, J. B. Reitan, F. Strand, T. Strand, F. Egil Skjeldestad. 1991. Pregnancy outcome in Norway after Chernobyl. *Biomedicine & Pharmacotherapy* 45 (6): 233–241.

Knudson, L. B. 1991. Legally induced abortions in Denmark after Chernobyl. *Biomedicine & Pharmacotherapy* 45: 229–231.

Lie, Rolv Terje. 1992. Birth defects in Norway by levels of external and food-based exposure to radiation from Chernobyl. *American Journal of Epidemiology* 136, no. 4 (August 15): 377–388.

Odlind, V. and A. Ericson. 1991. Incidence of legal abortion in Sweden after the Chernobyl accident. *Biomedicine & Pharmacotherapy* 45: 225–228.

Ramsay, C. N., P. M. Ellis, and H. Zealley. 1991. Down's syndrome in the Lothian region of Scotland—1978 to 1989. *Biomedicine & Pharmacotherapy* 45: 267–272.

Rich, Vera. 1992. Bielarus [sic]: Political fallout from Chernobyl. *Lancet* 339 (February 22): 484–485.

Sperling, K., J. Pelz, R. D. Wegner, I. Schulzke, E. Struck. 1991. Frequency of trisomy 21 in Germany before and after the Chernobyl accident. *Biomedicine & Pharmacotherapy* 45 (6): 255–262.

Spinelli, A., and J. F. Osborn. 1991. The effects of the Chernobyl explosion on induced abortion in Italy. *Biomedicine & Pharmacotherapy* 45: 243–247.

Summary of the Conference Results. 1996. Joint EC/IAEA/WHO International Conference: "One Decade After Chernobyl: Summing Up the Consequences of the Accident."

Technical Symposium [sic], Unofficial Summaries of. 1996. Joint EC/IAEA/WHO International Conference: "One Decade After Chernobyl: Summing Up the Consequences of the Accident."

Truth and Chernobyl. 1991. *New Scientist* 130, no. 1765 (April 20): Comment.

Webb, Jeremy. 1991. Chernobyl findings 'excessively optimistic.' *New Scientist,* June 1, 17.

Williams, Nigel, and Michael Balther. 1996. Chernobyl research becomes international growth industry. *Science* 272 (April 19): 355–356.

6

Should Lesbians Count as Infertile Couples? Antilesbian Discrimination in Assisted Reproduction

Julien S. Murphy

Assisted reproduction not only offers a variety of services for treating infertility, but includes some services useful for fertile women. One method of conception, physician-assisted insemination, can be helpful to fertile women who wish to use insemination as their preferred method of conception. Single women, lesbians, and lesbian couples are among those who rely on this method.[1] A few years ago, my partner and I began using assisted reproduction to conceive our first child, and it occurred to me as I wondered about lesbians' access to reproductive services, insurance coverage, and parenting rights for nonbirthing partners that there was little difference between us and the many infertile heterosexual couples for whom reproductive services were designed. While we lacked a medical reason for an infertility diagnosis, the similarities in the treatment plan and goal suggested that perhaps lesbian couples might be regarded as having a sort of "relational infertility" that could be said to accompany lesbian relationships.[2] Armed with a medical diagnosis, our reproductive concerns would be seen as legitimate. Our access to services would increase, they would be covered by insurance, and we would be granted the crown jewel of benefits afforded married heterosexual couples using donor insemination—parental rights for the nonbirthing partner. Despite similarities between our situation and that of those who are routinely diagnosed as infertile, infertility specialists do not regard as infertile lesbian couples who use physician-assisted insemination. But why not? Is it due to antilesbian discrimination in assisted reproduction? I began reading the medical and philosophical literature with the question in mind: Should lesbians count as infertile couples? It is a question of strategy with complex theoretical implications.

It may seem foolish to want a diagnosis of infertility, since medical diagnoses are rarely coveted. The fact is that accompanying an infertility diagnosis are some desirable privileges that lesbians currently lack. If lesbians want these privileges, perhaps we should be demanding the same diagnosis and trying to convince the medical establishment that lesbian couples seeking assisted reproduction are as infertile as male-female couples using assisted reproduction. As a result, we might have a good chance of obtaining the same privileges, and we would advance lesbian reproductive rights and increase the legitimacy of lesbian (and gay) families.

In this chapter, I focus on physician-assisted insemination. In order to evaluate the benefits of applying an infertility diagnosis to lesbian couples pursuing physician-assisted insemination with no known fertility impairment, I will describe this reproductive method within a lesbian and feminist context. Next, I identify three benefits of an infertility diagnosis accompanying assisted insemination for heterosexual couples, benefits that lesbians lack and that constitute forms of antilesbian discrimination. A review of arguments on physician-assisted insemination in the medical literature reveals common assumptions about lesbians and reproduction that reinforce all three forms of discrimination. Finally, I evaluate parity arguments in favor of applying an infertility diagnosis to lesbian couples. I argue that it is wise to be wary of certain tempting diagnostic strategies when searching for ways to address antilesbian discrimination in assisted reproduction.

Lesbian Insemination Practices

Lesbians who choose pregnancy without heterosexual intercourse use some form of (alternative) insemination. It has been estimated that as many as 10,000 children have been born to lesbians this way (Cohn 1992, 39).[3] Insemination is popular for many reasons. It can be significantly easier, quicker, and less costly than adopting a child (many adoption services discriminate against lesbian and gay clients). The most popular insemination is unassisted or self-insemination, whereby a woman is not aided by a health care professional. In the late 1970s, the feminist self-help movement promoted self-insemination for women wishing to reproduce on their own, and formal and informal sperm banks were established, some specifically for lesbians (Klein 1984). In engaging in self-insemination, usually at home, women directly seized control over conception, enhancing independence from men and a personal sense of freedom from heterosexual relationships and the medical establishment.

While unassisted reproduction represents an advance in reproductive freedom for lesbians and single women, the trend in the 1990s is to seek out fertility experts when self-insemination fails to achieve pregnancy. Increasingly, lesbians are pursuing assisted reproductive options (Hornstein 1984). Fertility clinics offer a range of reproductive techniques to increase the chances of pregnancy. They include vaginal as well as intrauterine insemination, hormone injections, ovarian hyperstimulation, egg harvesting for in vitro fertilization and embryo transfer, gamete intrafallopian tube transfer, and other techniques; eventually, human cloning may be possible. I wish to explore antilesbian discrimination in an early intervention, physician-assisted (vaginal or intrauterine) insemination.

Antilesbian Discrimination

Antilesbian discrimination in assisted reproduction occurs in at least three ways. The first is access. Some fertility clinics and physicians refuse to extend services to single women and lesbians (Robinson 1997). Restricted access to infertility services is not unique to the United States. It is common for fertility centers throughout Western countries to restrict physician-assisted inseminations to heterosexual couples, although there are exceptions (Knoppers and LeBris 1991; Daniels and Taylor 1993; Arnup 1994). Physicians regulate lesbian access not only to infertility techniques, but also to sperm banks (both physician-operated and others requiring physician authorization). Insurance plans that cover fertility services only if there is a diagnosis of infertility also regulate and discriminate against lesbians.[4] Even after one gains access to reproductive services, discrimination can still occur from nurse-clinicians, midwives, and other health care providers who may withhold information and support for lesbian patients or in other ways create a hostile atmosphere for lesbians.

The second form of discrimination already alluded to occurs in the use of infertility diagnosis and perceptions about the value and meaning of lesbians' requests for reproductive services. A lesbian couple in a committed relationship who seek out a physician for insemination services is not regarded as "infertile," unlike a married couple experiencing male infertility problems that require insemination by donor.

The diagnosis of infertility legitimizes the heterosexual couple as a "reproductive couple." In the minds of many, the diagnosis declares that the couple's desire to reproduce is not frivolous or optional (as might be thought of a single woman who wishes to reproduce on her own), but a necessity for the couple that is central to their very beings. Rare is the physician who regards a lesbian couple's desire to reproduce as similarly "necessary" to their life together. It is commonly assumed that lesbians will not reproduce and are not reproductive couples. This creates a vicious circle.

The diagnosis of infertility not only alters perceptions about couples' reproductive interests, it also provides heterosexuals with certain important benefits. Among them is the important benefit of paternity. In the case of a married heterosexual couple, if a sperm donor terminates paternity claims in advance, the husband of a woman having insemination is permitted to *adopt* any offspring that result from donor insemination. Many states have statutes that grant him *automatic* parental rights, provided that the insemination occurred in a doctor's office and that he has acknowledged in writing that he accepts the child as his own. Such statutory rules deem the consenting husband the legal father for all purposes. No adoption procedures, not even stepparent adoption procedures to examine his suitability to parent, are required (Andrews 1988; Chambers 1996). This is the third form of antilesbian discrimination associated with physician-assisted insemination. The nonbirthing partner in a lesbian couple does not have *any* automatic parental rights and commonly lacks any parental rights whatsoever.[5] Even in places where adoption is permitted, social workers and judges can, and occasionally do, insist on longer procedures or additional steps for gays and lesbians.[6]

Antilesbian discrimination practices in access, perception, and adoption are intimately linked. A review of the scholarly literature on the topic reveals that the main reason for refusing access to lesbians for assisted reproductive services is a failure to recognize lesbians as reproductive, hence as capable of having treatable fertility problems. It is assumed that lesbians are incapable of having families. Insurance companies, even in states mandating infertility coverage, can refuse to cover lesbians by claiming that fertility procedures are not "medically necessary" (Millsap 1996). Hence, conservative judges refuse to legitimize lesbian families by failing to grant adoption requests of nonbirthing partners. The standard assumption is that only heterosexuals *can* reproduce, which comes to mean that only heterosexuals *should* reproduce. If lesbians request medical assistance in reproductive technology, such requests are deemed special requests. These assumptions and their allegedly "special" status are borne out by the medical literature, albeit scanty and with very little discussion of physician-assisted insemination for lesbians.[7]

Physician-assisted insemination for lesbians was first discussed and sharply condemned in 1979 in letters to a British medical journal in response to an ethics committee report for the British Medical Association (Thomas 1979). Opponents argued that physicians would be unable to foresee possible trauma to the child from being raised in a nontraditional family (Cosgrove 1979), or from not knowing his or her father (Hatfield 1979), or that *the procedure itself was unwarranted lacking a clear medical condition* (emphasis added) (Wilson 1979). How we regard the reproductive interests of lesbians is, at the very least, a philosophical matter. Perceptions about lesbians' reproductive possibilities are often linked with assumptions about lesbian parenting, as well as about lesbians and infertility. Do lesbians "lack a clear medical condition" for infertility? There are striking similarities between lesbians choosing insemination and heterosexuals suffering from some form of male or female infertility that is treatable by donor sperm and insemination. The same treatment yields the same result, a pregnancy. Does that suggest that the "condition" is essentially the same? If so, then lesbians do possess "a clear medical condition."

One issue is whether lesbians are asking physicians to "fix a problem" (i.e., infertility) or to assist in fulfilling a possibility. This is an important philosophical issue, and yet there is very little discussion of lesbians and reproduction or even lesbian parenting in the philosophical literature.[8] If lesbians are asking physicians to assist in fulfilling a possibility, lesbians can be compared to single women desiring the same services and, presumably, *lacking a medical condition.* Conservatives champion this position and use it to restrict access to reproductive services for lesbians and single women. One opponent of physician-assisted lesbian insemination argued that it was more like psychotherapy than infertility medicine. Insemination could not be regarded as medical treatment in lesbian cases, he argued:

> *Treatment for what,* I ask myself. Infertility? No, it can't be that or it would not work. *Treatment for lesbian tendencies?* No, because the women are still lesbians after conception has occurred. *Treatment for the psychological strain of having "married" someone of the same sex and so being unable to conceive naturally?* This seems to me to be the only possible way in which the process can be regarded as "treatment."(Wilson 1979)

As the literature shows in an American case, it is assumed that physicians have a *duty* to treat "medical conditions," but addressing the fulfillment of reproductive possibilities is an optional matter. If lesbians are not perceived as individuals suffering from some treatable form of infertility, services can easily be denied.

Six years after D. H. Wilson's comments appeared, a highly publicized American case of physician-assisted lesbian insemination made its way into the *Archives of Internal Medicine* (Perkoff 1985). The lesbian patient was twenty-eight years old, in a monogamous interracial relationship for over a decade, and the couple wished to use an anonymous donor of the same race as the partner. The physician proceeded with several cycles of insemination after routine genetic screening and social work evaluations. The case came to public attention because of breaches in patient confidentiality that occurred when the patient sought treatment for unexplained abdominal pain and fever that developed during the third cycle of insemination. An anonymous letter to the local newspaper brought journalists to report the story. No doubt, the interracial aspect of their relationship and their choice of donor fueled the controversy, for the story was picked up by national wire services with follow-up stories in local and national newspapers. The breach of confidentiality had disastrous results for the lesbian couple and their physician: the inseminated woman did conceive, but a spontaneous abortion ensued at six weeks' gestation; the couple ceased inseminating; and because of the case and its publicity, their physician's job offer at a Catholic medical school was withdrawn. The lesbian couple was described as "deeply shaken" by the publicity and the repercussions for their physician (Perkoff 1985).

One analysis of this case identifies the breaches of patient confidentiality as unethical but finds no ethical principles violated by physician-assisted donor insemination for lesbian couples. The author did advise caution for physicians considering such requests (ibid.). John Fletcher, a bioethicist commenting on the case, explored whether or not the physician had a professional duty to perform donor insemination for the lesbian couple. He argued that physicians are not obliged to assist lesbian couples with insemination because there is no accepted medical condition requiring treatment, unlike cases of married couples with genetic disorders or male infertility. He concluded that physicians can freely refuse lesbians' requests and have no duty to refer patients to other physicians who might help them. Recognizing that such requests by lesbian couples are "not only 'unusual,' but a real break with traditional practices and the majority religious view on sexuality and parenthood," he urged the leaders of teaching hospitals to provide guidance on the issue (Fletcher 1985).

It is unfortunate that this case introduced American physicians to physician-assisted lesbian insemination because the severe repercussions for the physician involved cast an ominous light on the issue.

Fearful of publicity, no doubt many physicians would thereafter think long and hard about consenting to such a request by a couple who, because of their sexual orientation, might appear "unusual." If there is to be no discrimination based on sexual orientation, then lesbians have as much right to reproduce as anyone else. Fletcher's conservative analysis is equally unfortunate both because it refuses to recognize that lesbians have a right to these services and because it was the only response to the case in the medical literature. At the time, a meeting was convened at the university hospital by the chair of family medicine to discuss the case with the chief of staff, the chair of surgery, the obstetrician responsible for the sperm bank, and the physician who performed the insemination. Opinions about the appropriateness of donor insemination for the lesbian couple were divided.

Lacking good grounds to deny lesbians access to reproductive services, physicians should be compelled to provide reproductive care for lesbian patients or, in cases of religious conflict, refer lesbians to other physicians. Whether or not lesbians are regarded as having a right to assisted reproduction depends in part on whether or not lesbians are seen as assimilated within heterosexual society. Assimilation was an important screening factor in a more extensive discussion that occurred in Europe in response to a Brussels reproductive center (one of the first) offering donor insemination to lesbians (Brewaeys et al. 1989). In its program of twenty-seven lesbian couples who requested physician-assisted insemination from 1981 to 1988 and who underwent a screening program, twenty-one couples were accepted. Fourteen pregnancies occurred in the twenty-one couples, with four couples having a second child. Their requests for insemination were termed "special" requests, and careful screening and counseling procedures were adopted. The criteria for psychologically screening lesbian couples required inspection of the personal histories of both women, relational patterns of the couple, and an assessment of their understanding of parenting and desire to parent (ibid.). The reasons cited for refusing insemination services to six of the lesbian couples were: "unresolved problems in the family of origin, doubtful homosexual orientation, lack of acceptance of their homosexual identity, instability in the current relationship, intolerance of their social milieu towards homosexuality" (ibid.).

It is difficult to evaluate the degree of prejudice, if any, in the screening process without understanding the cultural environment for lesbians in Brussels. For instance, personal history included a positive homosexual self-image; and relational patterns looked for a high degree of social acceptance of the couple in their family of origin, their neighborhoods, and their wider social and professional lives. In the United

States, good candidates for parenting might not have family acceptance (particularly if either partner comes from a conservative background), may have prejudiced neighbors, and may not be "out" in their professional lives. The last condition, one could argue, is necessary for good parenting, but the others may fall outside a couple's control. In Y. Englert's opinion piece, the successful applicants were described as highly assimilated into heterosexual society, "surrounded by more heterosexual couples than homosexual ones and certainly could not be considered as part of a homosexual ghetto" (Englert 1994). It is doubtful that the same emphasis on cultural assimilation would be applied to a heterosexual couple from a racial, ethnic, religious, or economic minority. His reference to a "homosexual ghetto" is similarly disturbing. Most likely, the ghetto refers to a lack of heterosexuals, rather than to a lack of economic resources. While any screening mechanism is subjective and can allow prejudice, there was a high acceptance rate for lesbian couples at the Brussels center (twenty-one of twenty-seven couples in one study [Brewaeys et al. 1989] and fourteen of fifteen couples in another study [Englert 1994]). Perhaps this was a function of open-minded psychologists conducting the screening, a high degree of social acceptability for lesbians in Brussels, or a self-selecting applicant pool that sought out the Brussels center with their special requests because they felt they easily matched the criteria.

The link between one's perception of lesbian reproductive interests (whether they are seen as optional or fundamental) and lesbians' access to reproductive services has been demonstrated in this analysis of the medical literature. The same arguments are used to refuse lesbians the more complicated reproductive services.[9] There is little chance, then, that we can improve lesbian access without improving the perceptions about lesbians who wish to reproduce. To battle misperceptions, we must address the claim that lesbian requests for assisted reproduction are "special requests," for this is part of the conservative language that classifies gay civil rights as "special rights." Should we confront conservative perceptions by claiming that lesbians seeking assisted reproduction have medical conditions? Should we argue that lesbians should count as infertile couples? By this, we would mean that lesbians reproducing without intercourse have a treatable form of infertility—temporary or relational infertility—and therefore ought to be granted full access to fertility services. And would this position increase the chances that judges and others would legitimize lesbian and gay families and accept the diagnosis of temporary infertility, so lesbians could be perceived as capable of making families?

A Diagnostic Strategy

It would not be unreasonable to regard lesbians seeking assisted reproduction as temporarily infertile. The term "infertility" is one of the few relational terms in medicine. Medicine generally assumes that individuals alone have diseases or medical conditions and that the source of a medical problem cannot lie in the relation between the individuals. Infertility is an exception, for the reproductive couple may be the source of the problem. Much of the reproductive literature on this topic refers to diagnosing, treating, and counseling the infertile *couple*. Medical diagnostics allow for the possibility that the source of infertility, in some cases, may lie with the couple and not one of the individuals (Collins 1995; Seibel 1993; Daley et al. 1996). Given the relational aspect of fertility, two people may be fertile apart but infertile together. Infertility can be general or partner specific; it can be temporary or permanent. Approximately 5 to 10 percent of infertility in heterosexual couples is unexplained (Talbert 1992). Lesbians may well be fertile women but could be said to make infertile couples when assistance of insemination is required. By contrast with heterosexuals, lesbians are often not recognized as being in relationships. Medical literature on lesbian couples seeking insemination talks of "insemination in a lesbian" and never a lesbian *couple*. Also, lesbians needing only donor insemination and not other services such as in vitro fertilization are fertile and need only minor medical intervention. The clinical measure of infertility is biased as well. Infertility, as measured in the estimated 2.8 million infertile heterosexual couples who desire to reproduce, is the failure to conceive after one year of unprotected intercourse (Talbert 1992). Attempts to conceive are usually more numerous in noninseminating couples who incur no expense. It might seem reasonable, then, to extend the diagnosis of "temporary infertility," currently applied to some heterosexual patients, to lesbians seeking insemination services.[10]

If we opt to diagnose reproductive lesbians as temporarily infertile, will it change common perceptions of lesbians as nonreproductive, permanently infertile women? Perceptions of sterility are found not only in the medical literature, as mentioned above, but also in legislative arguments favoring state bans on gay marriage. In discussions of gay marriage, conservatives claim that the function of marriage is to raise families, hence gays and lesbians should not be allowed to marry because they cannot biologically reproduce together. The deviant status of lesbianism is associated with an assumed incapacity to bear children in lesbian relationships. Once it is pointed out that there already are gay and lesbian families, conservatives next raise concerns about the welfare of children in these families, even though there is extensive research demonstrating the well-being of children in gay and lesbian families (Golombok and Tasker 1994). Concerns next turn toward fatherhood, considered to be under siege by the practice of lesbian insemination. The belief that manhood is devalued by lesbian insemination, because it makes the role of men inconsequential, is found in discussions about lesbian insemination at the Brussels clinic. The opposition claimed that "the right to equal opportunities cannot entail the right for a woman to procreate without a man any more than the opposite scenario" and that "fatherhood" should not be "reduced to a gamete, or an *anonymous sperm cell*" (Shenfield 1994). These views reflect conservative positions on the family based on natural law philosophy. Lesbian reproduction does eliminate fatherhood in many instances and challenges common sexist assumptions that parents of both genders are necessary for good parenting. Feminists are among the advocates of lesbian motherhood. Advocates for physician-assisted insemination in lesbians argue that "as in other conflicting areas of women's rights it is the law that should adapt to the autonomy of women (where ethically acceptable) and not the other way around" (Englert 1994).

Some Considerations

The strategy of insisting on a diagnosis of temporary infertility for lesbians and lesbian couples enlists the authority of the medical establishment to legitimize the reproductive interests of lesbians. One can argue that such a strategy is a fair-minded demand for parity in reproductive rights because lesbians are analogous to heterosexuals who also seek donor insemination but who have a diagnosis of infertility. Further, the goal of greater access to reproductive services is an important one. Nonetheless, there are at least five good reasons to question this strategy.

First, in seeking parity with heterosexuals by demanding the same diagnosis, lesbians would be required to fit reproductive experiences into a heterosexual paradigm. Rather than struggling to confront or resist the dominant paradigms and broaden the models of reproduction, this strategy would encourage greater assimilation by lesbians into heterosexual categories, thereby forgetting important differences among these choices. Some of these differences are identified in a research study on why lesbians request physician-assisted insemination. In a study of fifteen lesbian couples at the Brussels reproductive clinic pursuing insemination from 1988 to 1993, the reasons for physician-assisted insemination were: refusal to sleep with a man and/or break the couple's fidelity (80 percent of the couples cited this reason); desire not to introduce a third party in the couple's project (53 percent); moral reluctance to deceive a male partner by engaging in sex without disclosing one's procreative intent (53 percent); and fear of contracting HIV disease (47 percent) (Englert 1994). There are other differences between lesbian and heterosexual practices of insemination worth considering. One interesting difference is that donor insemination is more likely to be disclosed to children in lesbian-headed than in heterosexual families. In studies of heterosexual couples undergoing donor insemination, most do not plan to tell the child (Klock, Jacob, and Maier 1994).

Second, granting the diagnosis of infertility to otherwise fertile lesbians extends the boundaries of what some call "compulsory motherhood" beyond heterosexuality, a regressive move. Social pressure for women to reproduce is powerful and pervasive. Since lesbians escape this expectation (once their sexual preference is made known), it would be foolish to recreate it. That said, we are in the midst of a lesbian baby boom. The days of lesbians being free of the expectation to reproduce are numbered. A diagnosis of infertility, by increasing access to services, would most likely create social expectations for lesbians to reproduce. This is partly because our society continues to socialize most girls to believe that motherhood should be their primary source of fulfillment in adult life. Also, once new medical technology, especially reproductive technology, becomes available, there is an impetus to use it, and it is widely marketed. While the assumption that lesbians cannot reproduce is oppressive, the assumption that women *must* reproduce is also problematic.

Third, a diagnosis of temporary infertility renders an aspect of lesbianism, once again, a medical condition. Little does it matter that initially it would be progressive gynecologists, not conservative psychiatrists, applying the new diagnosis. The inference would remain that there is something medically deficient about the reproductive bodies of lesbians because infertility is generally regarded as an abnormal condition. There is an important difference between lesbian couples and infertile heterosexual couples that we have not mentioned. While some heterosexual couples are regarded as infertile, the majority of heterosexual couples are not. A diagnosis of infertility for lesbian couples would suggest that lesbians are always infertile. In this way, a diagnosis of infertility would be more limiting for lesbians than for heterosexuals.

An important difference between heterosexual and lesbian couples seeking assisted reproduction is that the former can legally marry. Before arguments demanding parity in infertility diagnoses can be fully assessed, we must ask why some privileges accompany donor insemination in heterosexual couples. One important privilege, automatic paternity, applies only to legally married infertile couples. This suggests that paternity rights are grounded in a couple's marital status first and foremost and only secondarily in their infertility diagnosis. The assumption that heterosexuals are normally reproductive and thereby have access to infertility services may also be based on their marital status or at least their right to marry. If legal marriage is the fundamental basis for the benefits currently extended to heterosexual infertile couples, an infertility diagnosis for lesbian couples, barred from legal marriage, may not improve lesbians' access to services or nonbirthing partners' rights to be legal parents of children conceived by donor insemination.

Fourth, medical diagnostics further regulate people. Once lesbians are classified under a particular diagnosis, reproductive concerns shift from the bedroom to the physician's office, we become patients, and the formerly private matter of conception is taken over by the hegemonic practices of institutionalized medicine. Should we relinquish control over reproduction by seeking a diagnosis related to our relationships rather than to our bodies? Already, many who pursue assisted reproduction for infertility problems are subjected to regulatory practices. This regulation includes not only psychological screening, partially ideological in nature, as described above, but the chaotic, short timetables of fertility cycles, implantation procedures, and so forth, that take over normal schedules and priorities, as well as the displacement of the reproductive couple or individual by an extensive team of specialists. Since assisted insemination is a simple procedure, it is unlikely to develop the degree of medical control that may characterize more complex infertility procedures. Nonetheless, extending a diagnosis of infertility to otherwise fertile lesbian couples does further medicalize donor insemination. In addition, sperm banks define and regulate both donors and clients and have become a profitable industry. One researcher claims that since frozen sperm came into common use, "the results have been a major change in the manner that TDI [therapeutic donor insemination] is administered: costs have escalated, commercial sources of semen have expanded their distribution networks and the efficiency of fertilization through insemination has been diminished" (Shapiro 1994, 150).

While my strategy of extending an infertility diagnosis to lesbians has not been discussed in the literature, there has been discussion of the trend toward medicalization of donor insemination. Some who argue for self-insemination rather than the medicalization of donor insemination claim that medicalization offers "a measure of confidentiality," but also "involves intrusion and the possibility of some loss of control" (Wikler and Wikler 1991). Lack of access is cited as the most important problem with medicalization. This is interesting since increased access is a goal of my proposed strategy. If it is likely that increased medicalization leads to decreased access to donor insemination, as the Wiklers have found, then my strategy, accompanied by perhaps new or more scrutinized screening procedures, might prohibit access for more lesbians than it allows. In addition to decreased access, those who argue against the medicalization worry about the increased power it gives to doctors to determine who will and who will not reproduce. D. Wikler and N. Wikler argue against doctors' playing such a role in society and prefer that individuals retain full control over their reproductive choices. However, what they do not address is that where lesbians are driven away from fertility clinics and thereby prevented from achieving desired pregnancies, physicians also remain in control.

Those who oppose medicalization of donor insemination argue for the deregulation of donor insemination by encouraging self-insemination and demanding that restrictive practices by sperm banks, such as the requirement of physician authorization, be dismantled. Some such restrictions are especially problematic for lesbians. For instance, in some states, termination of paternity rights of a known donor is more easily accomplished if physician-assisted insemination is used. In light of my discussion, encouraging self-insemination over physician-assisted insemination should include teaching women the techniques for safe intrauterine insemination. There will be many who will still prefer physician assistance either because it is easier and more reliable and involves screening for sexually transmitted disease, or because medical and legal restrictions never get fully removed. Would an infertility diagnosis be advantageous for addressing discriminatory practices in assisted reproduction?

A final problem with my strategy is that to regard reproductive practices as "treatment for a medical impairment," rather than an elective procedure, is not liberating. A diagnosis of a medical impairment is often stressful. Many couples who currently fit the medical diagnosis of infertility have experienced great angst and frustration because of their inability to reproduce without assistance. They may feel an important function of their bodies has failed them. While lesbians may agonize over fertility matters (such as the inability to produce a child that is the genetic offspring of both women), frustration is largely over social barriers to assisted reproduction and prejudices against lesbian reproduction. By contrast, elective procedures are often liberating, particularly when they involve new medical options and do not presuppose a medical deficiency. Classification as an elective procedure would underscore the deliberate choice of pregnancies resulting from insemination.

If we regard assisted reproduction for lesbians as an elective procedure, we might reject the assumption that lesbian couples are similar to medically infertile heterosexual couples. Were it not for several significant points of dissimilarity, we might find more compelling similarities between lesbian couples and single women pursuing motherhood through assisted insemination. Single women pursuing donor insemination do not seek a diagnosis of infertility. For many, the option to reproduce outside of a relationship is liberating. But lesbian couples are reproducing *within* relationships. This is an important difference. Also, single women may choose assisted reproduction because there is no one with whom they wish to reproduce. Many lesbians long to be able to reproduce with their partners, a difference shared with infertile heterosexual couples.

Conclusion

Lesbian couples seeking physician-assisted insemination are disanalogous in important respects to both infertile heterosexual couples and single women. Yet there are two choices before us: to regard such services as medical treatment for an impairment (hence, a diagnosis of infertility) or as an elective procedure. Currently, lesbian couples are regarded as seeking the latter. In my experiences with assisted reproductive practices in both a large urban West Coast hospital and a small New England obstetric practice, I was surprised at nurses' responses to us. They were thrilled by the idea of lesbians reproducing in this manner. To them, it was a mark of women's liberation that two women need not marry or be involved with men in order to have children. For these nurses who routinely work with infertile couples, we were not "fixing a problem," we were choosing a new possibility. Would the freedom of this choice be diminished by an infertility diagnosis? Such a diagnosis could not avoid suggesting some impairment or deficiency in need of repair. If there are other ways to improve access to services and diminish discrimination, while still regarding assisted reproduction as an elective procedure, we would retain some measure of the freedom that currently frames the choice of assisted reproduction for lesbians.

Whether or not lesbians should count as infertile couples is a philosophical matter of political importance. Applying an infertility diagnosis to lesbians requesting physician-assisted insemination may normalize these reproductive practices for physicians and even win the support of insurance companies who rely on diagnoses as criteria for medical coverage to reimburse lesbians for insemination services. However, an infertility diagnosis may not greatly increase access to services or persuade adoption judges to approve adoption requests of nonbirthing partners if the process of obtaining the diagnosis fails to diminish antilesbian prejudice. Furthermore, convincing physicians that lesbian couples are "temporarily infertile" would be no easy task. This strategy would most likely be oppressive rather than liberating if any of the five political implications prove true. While the strategy to argue for parity for lesbians at the level of diagnosis may be quite tempting, we must resist it. We are better off seeking these benefits in other ways and steering clear of any infertility diagnosis.

I leave to my readers who have patiently followed this philosophical thought experiment and are wary of the problems I identify to devise better strategies so desperately needed to improve access to reproductive services, to alter perceptions of lesbian reproductive capacities, and to obtain parental rights for nonbirthing partners.

Notes

I am grateful to Becky Holmes and Timothy Murphy for comments on an earlier draft of this essay and to Barbara Katz Rothman for our conversation at the Women and Genetics in Contemporary Society Workshop on physician-assisted reproductive practices. I appreciate the support for this work provided by the Northwest Center for Research on Women and the Health Science Library at the University of Washington, Seattle, where I was a visiting scholar in 1995 and 1996.

1. Of course, some lesbians and single women do have infertility problems and use assisted reproduction services to address diagnosed infertility conditions.

2. I am grateful to Timothy Murphy for suggesting the phrase "relational infertility."

3. The accuracy of this number is difficult to evaluate since there is no record of the use of self-insemination, and some who use physician-assisted insemination may not identify themselves as lesbian fearing discrimination.

4. State regulation of infertility coverage by insurance companies is not uniform. Three states (Illinois, Massachusetts, and Rhode Island) require insurance companies to cover nonexperimental infertility treatments; seven states (Arkansas, Maryland, Hawaii, New York, Delaware, Ohio, and West Virginia) require infertility coverage in general and limited use of in vitro fertilization. In unregulated states, infertility coverage is quite variable. Low-cost early intervention options are usually covered (Gilbert 1996).

5. A few states (New Hampshire and Florida) prohibit any adoptions by gays and lesbians. Only three states (Massachusetts, Vermont, and New York)

and the District of Columbia have legalized same-sex second-parent adoption. It is permitted in progressive counties in eight other states (Alaska, California, Michigan, Minnesota, New Jersey, Oregon, Texas, and Washington) (Witt, Thomas, and Marcus 1995). In addition, there are favorable appellate decisions in Rhode Island, Colorado, and Illinois.

6. For instance, in some places, a full adoption procedure is required, while in others, stepparent or second-parent adoption, a shorter and less costly procedure, has been extended to gay and lesbian partners.

7. A review of the medical literature on lesbian insemination in *Medline* (1970–1996) shows some discussion of HIV risk, access, and prejudice (Shaw 1989, 135); and accounts of self-insemination (Hornstein 1984; Klein 1984), of lesbian childbearing (Zeidenstein 1990), and of child rearing (Golombok and Tasker 1994).

8. A review of the philosophical literature from *The Philosophers Index* (1970–1996) shows two articles on lesbian parenthood (Hanscombe 1983; Robson 1992) and only one article, a case conference, specifically about physician-assisted insemination in a lesbian couple in London (Forster et al. 1978).

9. See, for example, the discussion concerning a lesbian's request to undergo in vitro fertilization with ova harvested from her partner in Chan et al. 1993.

10. While it may seem more accurate to produce a diagnosis of "relational infertility" to all lesbian couples seeking assisted reproduction through insemination, it currently is not a term used in medicine.

References

Andrews, Lori B. 1988. Alternative reproduction and the law of adoption. In *Adoption Law and Practice*, ed. Joan H. Holinger. New York: Matthew Bender & Co.

Arnup, Katherine. 1994. Finding fathers: Artificial insemination, lesbians, and the law. *Canadian Journal of Women and the Law* 7 (1): 97–115.

Brewaeys, A., H. Olbrechts, P. Devroey, and A. C. Van Steirteghem. 1989. Counseling and selection of homosexual couples in fertility treatments. *Human Reproduction* 4 (7): 850–853.

Chambers, David L. 1996. What if? The legal consequences of marriage and the legal needs of lesbian and gay male couples. *Michigan Law Review* 95 (2): 447–491.

Chan, Connie S., Janis H. Fox, Richard A. McCormick, S. J., and Timothy F. Murphy. 1993. Lesbian motherhood and genetic choices. *Ethics and Behavior* 3 (2): 211–222.

Cohn, Bob. 1992. Gays under fire. *Newsweek*, September 14; 37–40.

Collins, J. A. 1995. A couple with infertility. *Journal of the American Medical Association* 274 (14): 1159–1164.

Cosgrove, I. M. 1979. AID for lesbians (letter). *British Medical Journal* 2: 495.

Daley, Jennifer, Thomas L. Delbanco, and Janet Walzer. 1996. A couple with infertility, one year later. *Journal of the American Medical Association* 275 (18): 1446.

Daniels, Ken, and Karyn Taylor. 1993. Formulating selection policies for assisted reproduction. *Social Science and Medicine* 37 (12): 1473–1480.

Englert, Y. 1994. Artificial insemination of single women and lesbian women with donor semen. *Human Reproduction* 9 (11): 1969–1977.

Fletcher, John C. 1985. Artificial insemination in lesbians: Ethical considerations. *Archives of Internal Medicine* 145: 419–420.

Forster, Jackie, Carola Haigh, Ian Kennedy, Anthony Parsons, Jennifer Pietroni, Rose Robertson, and Roger Higgs. 1978. Lesbian couples: Should help extend to AID? *Journal of Medical Ethics* 4: 91–95.

Gilbert, Bonny. 1996. Infertility and the ADA: Health insurance coverage for infertility treatment. *Defense Counsel Journal* 63 (1): 42–57.

Golombok, Susan, and Fiona Tasker. 1994. Donor insemination for single heterosexual and lesbian women: Issues concerning the welfare of the child. *Human Reproduction* 9 (11): 1972–1976.

Hanscombe, Gillian. 1983. The right to lesbian parenthood. *Journal of Medical Ethics* 9: 133–135.

Hatfield, F. E. S. 1979. AID for lesbians (letter). *British Medical Journal* 2: 669.

Hornstein, Francie. 1984. Children by donor insemination: A new choice for lesbians. In *Test-Tube Women: What Future for Motherhood?* ed. Rita Arditti, Renate Duelli Klein, and Shelley Minden. Boston: Pandora Press.

Klein, Renate Duelli. 1984. Doing it ourselves: Self-insemination. In *Test-Tube Women: What Future for Motherhood?* ed. Rita Arditti, Renate Duelli Klein, and Shelley Minden. Boston: Pandora Press.

Klock, Susan C, Mary Casey Jacob, and Donald Maier. 1994. A prospective study of donor insemination recipients: Secrecy, privacy, and disclosure. *Fertility and Sterility* 62 (3): 477–483.

Knoppers, B. M., and S. LeBris. 1991. Recent advances in medically assisted conception: Legal, ethical, and social issues. *American Journal of Law and Medicine* 17: 329–361.

Millsap, D'Andra. 1996. Sex, lies, and health insurance: Employer-provided health insurance coverage of abortion and infertility services and the ADA. *American Journal of Law and Medicine* 22 (1): 51–84.

Perkoff, Gerald T. 1985. Artificial insemination in a lesbian: A case analysis. *Archives of Internal Medicine* 145: 527–531.

Robinson, Bambi E. S. 1997. Birds do it. Bees do it. So why not single women and lesbians? *Bioethics* 3/4: 217–227.

Robson, Ruthann. 1992. Mother: The legal domestication of lesbian existence. *Hypatia* 7 (4): 172–185.

Seibel, Machelle M. 1993. Medical evaluation and treatment of the infertile couple. In *Technology and Infertility: Clinical, Psychosocial, Legal, and Ethical Aspects.* ed. Machelle M. Seibel, Ann A. Kiessling, Judith Bernstein, and Susan R. Levin. New York: Springer-Verlag.

Shapiro, Sander S. 1994. Therapeutic donor insemination. In *Progress in Infertility.* ed. S. J. Behrman, Grant W. Patton Jr., and Gary Holtz. Philadelphia: Lippincott-Ravens Publishers.

Shenfield, Françoise. 1994. Particular requests in donor insemination: Comments on the medical duty of care and the welfare of the child. *Human Reproduction* 9 (11): 1976–1977.

Talbert, Luther M. 1992. Overview of the diagnostic evaluation. In *Infertility: A Practical Guide for the Physician.* 3d ed., ed. M. G. Hammond and L. M. Talbert. Cambridge: Blackwell Scientific Publications.

Thomas, Michael J. G. 1979. AID for lesbians (letter). *British Medical Journal* 2: 495.

Wikler, D., and N. Wikler. 1991. Turkey-baster babies: The demedicalization of artificial insemination. *Milbank Quarterly* 69: 5–18.

Wilson, D. H. 1979. AID for lesbians (letter). *British Medical Journal* 2: 669.

Witt, Lynn, Sherry Thomas, and Eric Marcus. 1995. *OUT in All Directions: The Almanac of Gay and Lesbian America.* New York: Time Warner.

Zeidenstein, Laura. 1990. Gynecological and childbearing needs of lesbians. *Journal of Nurse-Midwifery* 35 (1): 10–18.

7

Equality, Autonomy, and Feminist Bioethics

Elisabeth Boetzkes

P hilosophical discussions of women's reproductive autonomy were once characterized by attempts to identify and balance rights or by the employment of relevant analogies (Overall 1987; Thomson 1971; Warren 1984). Currently, however, feminist philosophers are focusing on the relation among embodiment, reproduction, and women's self-identities. Typical of the current approach are both Catriona Mackenzie's account of pregnant embodiment (Mackenzie 1992) and Drucilla Cornell's treatment of equality (Cornell 1995). Mackenzie reformulates women's reproductive autonomy, arguing that our concept of autonomy must accommodate the woman's unavoidable exercise of moral agency and the unique bodily perspective generated by the phenomenology of pregnancy. The fact of pregnant embodiment, she argues, implicates a woman's self-identity, thus bodily autonomy cannot simply be understood as having control over what happens to our bodies. It is our capacity for moral self-constitution that is in need of protection when we demand reproductive autonomy.

Cornell also recognizes the link between reproductive autonomy and women's selfhood. Arguing that women's equal worth must be recognized and accommodated, she identifies several conditions for the possibility of personal individuation, which cannot be met if women are denied reproductive autonomy. The projection of the self's meaning, which is particularly necessary both for bodily integrity and for personal individuation, is incompatible with restricting reproductive freedom.

While acknowledging that the link among reproductive autonomy, self-constitution, and equality supports women's decisional privilege in reproductive matters, I will argue in this chapter that women's control over (what Cornell calls) the "imaginary domain" requires a public corrective to current female stereotypes and that this in turn may justify a legal response restricting the practices of sex selection and contract motherhood.

Reconceptualizing Autonomy

In "Abortion and Embodiment," Catriona Mackenzie argues for a reconceptualization of reproductive autonomy, one in which the responsibility of the pregnant woman is recognized in its full significance. Both women and men, as gamete providers, are causally responsible for a pregnancy, but there is a profound asymmetry between this causal responsibility and the further responsibility carried by the pregnant woman. Mackenzie distinguishes causal from moral responsibility and, within moral responsibility, decisional from parental. Once pregnant, she argues, a woman cannot avoid moral responsibility, for she must make the decision whether or not there will be a child of hers brought into the world, and she must make this decision in light of all other affected persons toward whom she bears responsibilities of other sorts. Her responsibility, unlike a man's, is unavoidable. She must decide whether or not to bear a child. In this sense, the exercise of moral agency and the assumption of moral responsibility are unavoidable for the pregnant woman.

Furthermore, the exercise of decisional responsibility by the pregnant woman deeply implicates her self-identity. For exercising decisional responsibility is simultaneously deciding whether or not to become a parent (perhaps again). The woman's decision is not just about the fetus. It is about what she is and what she will become. In deciding whether or not to bear a child the woman will engage with both personal and social meanings of pregnancy and parenthood, and she will consider whether she wishes to transform herself according to, and be judged by, those meanings. For, as we know, motherhood may invite censure or approval, depending upon factors such as age and marital and socioeconomic status (MacIntyre 1976; McCormack 1989). Motherhood may be perceived as an avenue of personal achievement (Morgan 1989), or as a daunting social burden (Held 1993). Is she in a position—socially, economically, psychologically—to care for a child, or even to view herself as a parent? What moral self-understanding is compatible with the decision she will make, and will her decision integrate or fragment her as a moral being? The moral responsibility of the pregnant woman is unavoidable and profound, and it engages her actively in a dialogue between the world of public meaning and her own self-understanding.

The challenge to the self-understanding of the pregnant woman that is posed by this dialogue is intensified by the phenomenological challenge to her subjectivity that pregnant embodiment poses. For as Mackenzie and others have pointed out,[1] for the pregnant woman, the distinction between self and other requires revision over time, as both physically and psychologically the woman perceives the fetus as both a part of her and yet apart.

> The foetus is not simply an entity extrinsic to her which happens to be developing inside her body and which she desires either to remove or to allow to develop. It is a being, both inseparable and yet separate from her, both part of and yet soon to be independent from her, whose existence calls into question her own present and future identity. (Mackenzie 1992, 148)

While pregnant and because she is pregnant, therefore, a woman faces a unique challenge to her self-identity, arising both phenomenologically and morally, and mediated culturally by the meanings of pregnancy and maternity. Through these meanings, she must scrutinize her view of herself and attempt to arrive at an integrated self-understanding and a compatible set of actions. Pregnancy is, thus, a profound opportunity for self-determination.

Mackenzie is not the only feminist arguing for such a broadened notion of autonomy. In *Self, Society, and Personal Choice,* Diana Meyers (1989) gives an account of autonomy that can situate Mackenzie's insight into the significance of pregnant embodiment. Rejecting the view that achieving autonomy is a private project of a liberated, self-monitoring inner self, she combines the recognition that selves are socially constructed with the argument that autonomy as a set of skills is nevertheless attainable and meaningful. The skills in question—self-discovery, self-definition, and self-direction—cannot be developed or exercised privately, for an important part of developing competency as an autonomous person is appropriating and testing out models of who and how to be, models available only in the public world. Achieving autonomy, on this view, must be a cooperative venture, one in which the person practicing the skills of autonomy benefits from having available a range of models to "try out." However, Meyers also notes that women's autonomy-competency is significantly hampered by enforced models of heteronomous altruism and the lack of socializing encouragement to try out various personae. If women are to have a chance at achieving full autonomy, we have a responsibility to make available a range of possible identities among which women attempting to self-define and self-direct may maneuver. Negative, rigid, or demeaning models of identity hamper the project of autonomy as self-determination.

Combining Meyers's analysis of autonomy with Mackenzie's analysis of pregnant embodiment and the responsibility of the pregnant woman engendered by it leads us to expand our understanding of what reproductive autonomy must include. First, we must recognize that the pregnant woman is not passive, but unavoidably an active moral agent. For,

> if causal and decision responsibility are inseparable for women, then pregnancy cannot be thought of as a merely "natural" event which just happens to women and in relation to which they are passive. [P]regnancy is never simply a biological process, it is always an active process of shaping for oneself a bodily and moral perspective. (Mackenzie 1982, 141–142)

Second, since in exercising responsibility the pregnant woman necessarily encounters the world of public meaning and shapes herself in response to it, pregnancy is a profound opportunity for self-determination. Viewing reproductive autonomy as merely a matter of respecting bodily privacy misses the significance of pregnant embodiment for the shaping of a moral self. To the extent that respecting autonomy involves recognizing the importance to persons of shaping and acting from a chosen moral perspective, respecting the autonomy of the pregnant woman involves much more than not invading her bodily boundaries. Furthermore, incursions into those boundaries must be recognized as not simply compromises of bodily autonomy, but compromises of moral agency and moral self-identity.

Third, the foregoing analysis suggests there is a public responsibility not just to acknowledge the "thick" nature of reproductive autonomy, but to play an active role in facilitating it. Since the pregnant woman must engage in a dialogue with public meaning, we must strive to make available meanings of pregnancy, maternity, and parenthood that are accessible, realistic, and compatible with respect for women as persons. The current cultural meanings of motherhood are, of course, diverse, and their integration into a given woman's identity distinctive; however, as feminists have argued, maternity is commonly structured in a depersonalizing or coercive manner. Pregnant women are represented (in law, medicine, and more broadly) as fetal containers (Annas 1986); models of compulsory altruism abound (Strickling 1988); maternal labor is trivialized or dismissed (Rothman 1989); or, alternatively, maternity is romanticized and idealized (Morgan 1989). Respecting the autonomy of the pregnant woman, then, must be a social project, not just involving a legal recognition of, for example, abortion rights, but developing realistic, adequate models of maternity and rejecting those that are demeaning or coercive. To acknowledge the importance of such models is to recognize that the project of autonomy and moral self-constitution is not a private project.

Reconceptualizing Equality

Drucilla Cornell also recognizes that the argument from privacy for reproductive autonomy is inadequate. In a lengthy defense of abortion rights, Cornell argues that the wrong of limiting reproductive autonomy is not that it invades a protected area, but that it undermines the process of personal individuation, by "placing the woman's body in the hands and imaginings of others who would deny her coherence by separating her womb from her self" (Cornell 1995, 38). The struggle is identified as one over the control of meaning.[2]

Like Mackenzie's revision of autonomy, Cornell's reconceptualization of equality focuses on the importance of public meaning. For Cornell, the demand for equality is the demand for an equal chance at becoming a person, a task requiring that three conditions be met: bodily integrity, access to symbolic forms adequate for the development of the linguistic skills necessary for individuation, and the protection of the imaginary domain. Taking a psychoanalytic approach, Cornell argues that becoming a person is importantly an activity of the imagination, a projection of various images of ourselves as somethings, a projection that must be acknowledged and affirmed by others. Thus becoming a person involves symbolic interplay between our imagination and "symbolic others," other individuals capable of recognizing and providing the meanings through which we project ourselves. Demanding equality is demanding that citizens be recognized as equally worthy of this attempt to become persons and that the basic conditions necessary for that task be provided.[3] As we shall see, the demand for equality will place restrictions upon the world of public meaning.

Each of these three conditions of personhood relies on the symbolic. Providing for the acquisition of adequate linguistic skills amounts to providing the means to learn a language, obviously fundamental to conceiving and reconceiving of ourselves. Bodily integrity involves more than just keeping the body intact. It is a matter of projecting an image of ourselves as whole and having that image "work." And finally, projecting an image of ourselves as whole somethings, individuating and identifying ourselves and eliciting recognition by others, is a symbolic activity, an activity of imaginative projection.

In this imaginary domain, we are vulnerable. We must depend upon symbolic others for ego confirmation (the acknowledgment of our projected images), and we are limited by the images available and the degree of maneuverability allowed. For example, as Cornell persuasively argues in relation to both abortion and pornography, at issue is whose meaning our bodies should have. Part of the injury of loss of reproductive control, or imposed exposure to pornographic representation, is the imposition of another's meaning on one's own body or sex. So, says Cornell,

> The right to abortion should not be understood as the right to choose an abortion, but as the right to realize the legitimacy of the individual woman's projections of her own bodily integrity, consistent with her imagination of herself at the time that she chooses to terminate her pregnancy. (Cornell 1995, 53)

And, again,

> When I have thrust upon me a fantasy of my body that completely undermines my own imaginary projection of bodily integrity, I am at that moment harmed because it undermines my ability to imagine myself as a person worthy of happiness, whose minimum conditions of individuation deserve to be protected equally. (Ibid., 150)

But the threat to the individual's imaginary domain is not just from imposed images. When the imposed images are degrading, the personalizing project is seriously compromised. Again speaking of pornography, Cornell says, "it is not the simple fact of our confrontation with images of ourself inconsistent with our own imaginary but a vision of ourselves as not worthy of the equivalent chance of personhood that is at issue" (ibid., 154).

Women are particularly disadvantaged by both the imposition and the downgrading of images, according to Cornell, because of the role being sexual plays in the development of the self. Being sexual, she claims, is so fundamental to being human that we unconsciously assume a sexual identity, a sexual imago, prior to embarking on the project of becoming a person. Through this internalized sexual imago, we individuate and shape ourselves as whole bodies and as persons. However, in the case of women, sexual identity has been encoded and symbolically enforced as feminine gender, thereby rendering the option of appropriate sexuality unobtainable and replacing it with (what Cornell calls) the "masquerade of femininity." Furthermore, the feminine, as encoded and enforced, has been downgraded, so that women are not only denied maneuverability around sexuality, but denied self-respect, since the feminine imago that we have internalized belittles our sexual difference. In response to such problems, Cornell seeks "to restore the feminine within sexual difference" by demanding equivalent opportunity to become persons. And an important part of this project is to insist upon what she calls a "degradation prohibition."

Although Cornell advocates a broad freedom within which individuals may explore and construct their personhood, her analysis supports limiting the imposition of degrading images. She explains,

> By degradation, I mean a literal "grading down" because of one's sex or sexuality. By a "grading down," I mean that one has been "graded" as unworthy of personhood, or at least as a lesser form of being. The treatment of a person as a "dumb-ass woman" or a "stupid fag" violates the degradation prohibition because it creates hierarchical gradations of sexual difference that scar some of us as less than persons worthy of happiness. (Ibid., 10)

The purpose of the degradation prohibition is not to impose alternative images, but to protect self-respect and to make space for the exploration and development of nonhierarchical self-understandings compatible with the power and creativity of personhood.

The state is doubly implicated in this activity. According to Cornell, the liberal state has a basic responsibility to treat citizens equally. If the opportunity to become a person is a primary social good, then the deservedness of all citizens to pursue it must be protected. The state's role here is twofold: to protect bodily integrity; and to protect self-respect, which is instrumental to the possibility of personhood, as she characterizes it. Protecting bodily integrity, as we have seen, involves more than just protecting citizens from injury. It includes freeing citizens from the imposition of bodily images incompatible with personhood (for example, the body as fetal container or sexual object). Further, the state must protect the imaginary domain, which is the seat of the pursuit of personhood, by exercising the degradation prohibition.

These activities would be incumbent on the state even if it were a detached mediator of equality. However, Cornell argues, the state is also a symbolic other, implicated in the construction of selves, for it is clear that legal representations of selves can facilitate or limit the development of personhood. She says,

> If we [accept] . . . the role of the Other in constituting a person, we can begin to think of a legal system as a symbolic Other; a system that does not merely recognize, but constitutes and confirms who is to be valued,

who is to matter. . . . Such an understanding of the legal system as "active", as a symbolic Other, validates a feminist claim for legal reform. It allows for a fuller appreciation of how the denial of legal and social symbolization can be so significant to whoever is confirmed as a self, and, in that sense, guarantees what I have called the minimum conditions of individuation. (Ibid., 42)

Skeptical doubts about the role of the state will be taken up later. But for now, let us explore the implications of the revised accounts of autonomy and equality for the practices of sex selection and contract pregnancy.

From Mackenzie's revisioning of autonomy and from Cornell's argument for equality, we see that achieving autonomy and equality depends upon support from the world of public meaning. If the pregnant woman explores and creates a self-understanding as she exercises her decisional responsibility, she is either limited or facilitated by the meanings available to her. Likewise, as women attempt to individuate, we may be victims of an impoverished public imaginary, or one in which the relevant meanings are, as Cornell says, "encoded and enforced." It is a commonplace that our reproductive practices are a major source of the meaning of being female, and, if, as I have claimed, the models of pregnancy and maternity are inadequate or coercive, a corrective to the public "maternal imago" is necessary. Certain reproductive practices, among them sex selection and contract motherhood, pose a particular threat to the possibility of autonomy and equality for pregnant women—indeed, for women in general. For the symbolism of sex selection, in a patriarchal society, proclaims that females are less valuable than males. In Cornell's terms, it is degrading. And the symbolism of contract pregnancy suggests that the personal integration and autonomous moral agency of the contract mother can be traded away. Both meanings diminish the possibility of full autonomy and equality for women, and, I will argue, neither is justifiable on the grounds of necessary benefit or political liberty.

Contract Motherhood

There are numerous criticisms of—and considerable disagreement about—contract motherhood; here I want only to address the autonomy of the gestational woman and the symbolism of the practice. Early formulations of the symbolic objection, such as those by George Annas (1986) and Christine Overall (1987), focused on the misrepresentation of the pregnant woman as inferior to her fetus and the commodification of women and children. However, in light of the link Mackenzie reveals among self-constitution, pregnant embodiment, and autonomy, the symbolism of contract pregnancy takes on a deeper, and more troubling, significance. For if being autonomous for the pregnant woman means engaging in an ongoing dialectic among the experience of pregnant embodiment, an examination of the cultural meaning of pregnancy and maternity, and the shaping of a personal and bodily perspective, then contractual restrictions actually preclude autonomous choice. The preconception contract binds a woman to a particular self-understanding, uninformed by the pregnancy she will undergo and compromised by the dominant, heteronomous, or romanticized public images of maternity.

But must contract pregnancy actually and symbolically limit women's autonomy? Some feminists have argued that autonomy is better served by keeping open the option of pregnancy contracts but ensuring their fairness and insisting upon a "change of heart" period, thus recognizing the significance of gestation. Rosemarie Tong, for instance, in a thorough discussion of the objections to contract pregnancy and feminist responses to them, concludes that the practice can be made consistent with feminist concerns about women's oppression by bringing it under the umbrella of the adoption model (Tong 1995). State surrogacy boards, she argues, could ensure fair contracts with provision for women to keep their babies, if they wish. Thus both reproductive and contractual autonomy are expanded, without the harms envisioned by both Marxist and radical feminists.

Prima facie, Tong's proposal gains support from Mackenzie's analysis. For if, as Mackenzie argues, to be autonomous the pregnant woman requires space for the self-constituting dialogue, knowing that she has that space in the form of contractual protections apparently reduces the threat to her autonomy. Furthermore, if the possibilities for self-constitution are enhanced by the range of alternative self-understandings available in the world of public meaning, a positive model of the contracting gestator might in fact enhance her autonomy, as Tong has suggested.

However, I have doubts about both the subsumption of contract pregnancy under the adoption model and the benefits of publicly endorsing contract pregnancy. First, would the "change of heart" option be seen as a true option by the gestating woman? Within the practice of adoption, the legitimacy of the gestating mother's claim to her child is never in question. At the same time, the claim of the adopting couple is seen as relatively weak and certainly not as sufficient for a grievance against the birth mother. While their disappointment may be acknowledged in a "change of heart" situation, they are hardly viewed as victims of an injustice. Feelings around contract pregnancy, however, are quite different. First, there is almost always a genetic connection to one or both social parent(s). This connection has been seen as strong enough to cast doubt on the claims of contract birth mothers. Second, the fact that there was an understanding prior to conception that the child would be handed over generates a moral expectation that is absent in the adoption case. These perceptions will be internalized by the contract mother, thus providing both internal and external pressure to forfeit the change of heart provision. Finally, where there is a brokerage system in place, even a disinterested government agency, there is a degree of awkwardness in "upsetting the applecart" by changing one's mind at the end. These factors, I submit, would exert pressure on the gestating woman, precluding her full participation in the process of self-constitution advocated by Mackenzie.

What about the argument that contract pregnancy expands women's options, thereby serving autonomy rather than limiting it? While this argument initially seems to gain support from the analysis of autonomy we have given, further scrutiny shows it to be problematic. For the putative benefit of expanded options must be weighed against the pervasiveness of an encoded femininity that already promotes reproductive labor as appropriate for women and endorses exaggerated female altruism. If we, as a group, institutionalize contract pregnancy, we simultaneously reinforce these stereotypes, thus limiting women's self-respect and possibilities for self-definition. Overall has expressed this concern in response to the Ontario Law Reform Commission's endorsement of contractual provisions such as those suggested by Tong. Overall observes,

> The legalization of contract motherhood would present reproduction for money as an acceptable, even desirable, aspect of women's place in Canadian society. But this path is incompatible with the vision of women as equal, autonomous, and valued members of this culture. . . . Is this a direction for female employment that should be encouraged? Should girls be urged to consider contract motherhood as a possible career? Does this practice promise a positive future for Canadian women? (Overall 1993, 131–132)

If the contract mother's pursuit of autonomy is undermined by either the decisional prematurity threatened (though, on Tong's model, not enforced) or by the contract and the paucity and inappropriateness of the available models of maternity, tolerance of the practice renders it problematic. Our silence about the discriminatory potential of the practice (for it will surely legitimize racist and ableist preferences) and its connection to degrading stereotypes intensifies symbolic harm, further compromising both autonomy and equality.

But is this concern about stereotypes and symbolic harms justified, and is the limitation of women's contractual freedom compatible with liberal politics? Richard Arneson has recently questioned both, discounting the symbolic objection and dismissing it as a form of legal moralism. Arneson argues that since the practice of contract motherhood is very much a minority practice, it will not affect the broader social meaning of reproduction (Arneson 1992). Furthermore, he argues, in a liberal society, nonharmful practices cannot be prohibited because their symbolism is distasteful to some members.

However, if reproductive autonomy for the pregnant woman requires ongoing decisional privilege as well as the availability of adequate models of maternity, Arneson is mistaken about the symbolic innocence of contract motherhood. Even if the practice of contract motherhood alone did not shape the public imaginary, it exacerbates an already common stereotype that discounts the importance of women's reproductive autonomy and is inordinately preoccupied with our reproductive effectiveness.[4] Furthermore, if the meaning of contract motherhood threatens women's personal individuation, Arneson cannot claim that contract motherhood is a defensible lifestyle option in a liberal state. For, if Cornell is right that the state has a responsibility to provide equal conditions for personal individuation, then it would follow that the state must act against practices that embed differential evaluations of social groups. Autonomy and equality are not optional values under liberalism.

Sex Selection

Like contract pregnancy, the practice of sex selection is discriminatory. If the conditions for autonomy include the availability of adequate models for female self-understanding, and if the conditions for equality include respect for the equal worth of individuals and groups as represented in the imaginary domain, then sex selection undermines both autonomy and equality. Feminist analyses of sex selection set the practice in the context of the transhistorical and transcultural devaluation of women, demonstrating that the practice of sex selection is not gender neutral in its distributive outcome (Williamson 1976; Steinbacher 1981). Within this political analysis, however, feminists have had difficulty launching a criticism of the practice that does not depend upon predictions about social futures that can be challenged (Warren 1985). Applying the present analysis, however, provides a foothold for a moral criticism of sex selection, without either having to make questionable, futuristic causal claims or engaging in the abortion debate. For if the symbolism of sex selection—which denigrates and devalues the female—renders women's personal individuation problematic, the practice is seriously morally and politically objectionable.

The practice of sex selection formally threatens autonomy and equality by representing females as having less value than males (Boetzkes 1994). Being undervalued as female precludes adequate personal individuation, since it jeopardizes the genesis of an appropriate self-image from the materials in the public imaginary. While women as women are publicly devalued, our task of becoming persons is rendered problematic.

However, being undervalued as female also jeopardizes self-esteem, which, as both Cornell and Meyers have argued, hampers the individual's capacity for individuation and substantive autonomy-competency, respectively. Without adequate self-esteem, the process of self-definition reduces to the appropriation of heteronomous models. If sex selection pronounces females inferior to males, the compromise of self-esteem is perpetuated, and the public imaginary is distorted, which in turn undermines both personal individuation and the development of autonomy. Again the state, as the protector of equality, cannot remain silent about this without abdicating its political responsibility.

But just as we asked whether allowing contract pregnancy might in fact serve women's autonomy and equality better than disallowing it, we must weigh the loss to reproductive freedom by rejecting the practice of sex selection. There are several arguments in favor of allowing access to sex-selection technologies. Some deny that sex selection is sexist and argue that, in Canada at least, there is no justification for limiting sex-selection technologies. Unless sex selection is intrinsically or instrumentally harmful, ruling out sex selection as a reproductive choice is indefensible. This view is seemingly supported by the finding of Canada's Royal Commission on New Reproductive Technologies (1993) that most Canadians surveyed are not sexist in the expression of their preferences for children.

However, such a position seems to me both naive and smug. (Our culture has advanced beyond sexism!) First, it is unclear whether sexist bias would be acknowledged or even detected in the instruments used by the commission. Second, even if individual Canadians were not sexist, the culture still suffers from systemic sexism, and the gains in equality women have made are too fragile to withstand official endorsement of sex as a sound reason for preferring one child over another. Denying individual Canadians that choice seems less a violation of individual autonomy than a cautious policy decision.

Withholding sex-selection technology, however, may have more seri-ous implications for women than simply the removal of an option. For, indeed, some women's refusal to participate in sex selection may en-danger themselves by alienating partners and their families. In such cases, are the gains to women's self-determination and equality real, and do they outweigh the losses to women's well-being?

There can be no general answer to this question, for whether the re-moval of the sex-selection choice is dangerous for women or strength-ens our equality by a public declaration of intolerance for sexism de-pends upon the social situation of the women in question. Where women have adequate economic and social protections and alterna-tives to compliance with partner and family, the symbolic gains proba-bly win. I suggest that in Canada the balance tips toward denying sex selection, and in this I reflect the attitude of many Canadian women most intimately affected, namely, members of the Indo-Canadian com-munities targeted by U.S. entrepreneurs seeking to create a market for sex-selection technology (Papp 1991).

Nevertheless, feminists must view with caution any call for repro-ductive restrictions. Cornell, herself commenting on the feminist di-lemma over sex selection, asks:

> What does it mean to allow abortion on the basis of sex under the protec-tion of the rubric of the right to abortion if it denies equivalent value to the feminine within sexual difference? My preferred solution is to criticize all exceptions to anti-abortion laws that turn on the hierarchies which evaluate whose body is to matter. This solution does mean that in a choice between protecting the right to abortion on demand, and legally restrict-ing the right in order to curb the practice of aborting female fetuses as a violation of the equivalent value of the feminine "sex", I would choose the former. This is, in part, because the resymbolization of women as sub-jects of rights is for me one prong of the overall challenge to the devalua-tion of the feminine within sexual difference. (Cornell 1995, 87)

Thus, for Cornell, the gains made for women's equality by unequivocal representation of us as rights-holders take priority over the symbolic denunciation of sex selection, since implicit in the latter is approval of limiting women's rights.

However, I disagree with Cornell on three points. First, it is an open question which strategy is more likely to strengthen the perception of women as subjects of rights. If sex selection thrives and is endorsed, the devaluation of the female will continue. Second, Cornell is discuss-ing sex selection in a context in which abortion rights are limited and being eroded. In Canada the situation is different (although with the rise of the populist Reform Party as the federal opposition, Canadians may soon be treading the American path). My concern, therefore, arises not in the context of the abortion debate, but instead in the attempt to draw up policy for offering or limiting new reproductive options, such as preconception sex selection and sex selection during in vitro fertilization (IVF). Finally, although, like Cornell, I endorse the creation of imaginative space within which the feminine can be reclaimed and reconstructed, I also recognize as a feminist that gains for individual women should not be made at the cost of exacerbating the oppression of women as a group. I would therefore not see as normative a woman's projected self constituted as "bearer of a male child" or "bearer of an able-bodied child." Just as the degradation prohibition must limit the effect on the public domain of the stereotypes of sex and sexuality, we can legitimately limit the effect of sexist or ableist preferences, even when it is women who self-define by means of them. I therefore see the arguments from autonomy and equality supporting a moratorium on sex selection as well as contract pregnancy.

Legal Remedies and Objections to Them

I began this chapter by claiming that the persuasive revisioning of the autonomy and equality of the pregnant woman may justify restrictive legal responses to the practices of contract pregnancy and sex selection. It appears that both practices limit women's autonomy through the reinforcement of degrading and/or stereotypical representations. Furthermore, the imposition of such representations denies women the equivalent chance to become persons. Both the threat to autonomy and that to equality are the state's business. Where disagreement exists about the harms to women's interests (understood traditionally) posed by practices such as sex selection and contract pregnancy, state permissiveness seems acceptable. However, when the connections among public symbolism, personal autonomy, and equality are exposed, the political argument for state involvement is strong. Autonomy and equality are central political values, to be compromised only where competing values are weightier. Since it is doubtful whether there is a positive right to reproductive assistance or to accessing practices or services that contribute to discrimination (Overall 1993), the demand for such "reproductive alternatives" as pregnancy contracts and sex-selection technology cannot outweigh the objections to them. It is therefore appropriate to expect the state, as a major contributor to and protector of the public imaginary, to declare such alternatives incompatible with the projects of political equality and personal autonomy. For the state, explicitly in law and implicitly in policy, authoritatively imposes or approves social categories, ranks individuals, and approves social relations. As Rae Langton (1993) has pointed out, legal utterances such as "blacks are not permitted to vote" rank blacks and legitimate discriminatory behavior, as well as depriving blacks of important powers. Such proclamations effectively subordinate. Further, state policies (such as population or education policies) may function in the same way. By contrast, upholding laws against racial or sexual discrimination, and proposing remedial or equitable policies, affirms equality and the right to personhood described by Cornell. Reproductive policies, like policies on race or sex, affirm or deny equality.

Critics, however, are bound to object to this suggestion on a number of grounds. First, does not the proposal to ban contract pregnancy and sex selection simply amount to legal moralism, the enforcement of positive morality by legal means? Philosophers such as H. L. A. Hart (1963) and Joel Feinberg (1984–1988) have long since debunked legal moralism and shown it to be incompatible with liberal politics. In suggesting that the state must protect the public imaginary, am I not invoking the slide into widespread censorship and, indeed, the thought police?

Like Cornell, I would argue that such proposals as the degradation prohibition, based as they are on the call for equality, are perfectly compatible with, indeed demanded by, liberal jurisprudence. Requiring the state to protect an equivalent right to bodily integrity and to the conditions for personhood is not the same as imposing a positive morality. Whatever moral content such protections have is shared among all attempts at a democratic polity. Hart and Feinberg are targeting "victimless crimes"; the victims of stereotyping and denigration are real enough, as are the consequences to their well-being. While the route from degradation to harm may be circuitous, it has recently been acknowledged as deserving of legal remedy by the Supreme Court of Canada in the famous cases *R. v. Keegstra* (1990) and *R. v. Butler* (1992), in which the Court upheld limits on racial propaganda and pornography, respectively. The Court adopted what Catharine MacKinnon has called a "harms-based equality approach" and explicitly distanced itself from both legal moralism and the offense principle.

But even if legally targeting degrading representations or practices that promote them is not moralistic, it may still find disfavor in feminist circles. Feminist skepticism about legal remedies is widespread and various. Some critics dislike the appeal to rights, which is oppositional and downgrades the importance of relationships (Menon 1993); some argue that the legal approach deflects attention and energy from more important social remedies (Fudge 1990); still others wonder whether institutions that have been vehicles of oppression can realistically be expected to be effective in its reversal (Smart 1989).

While all these objections must be taken seriously, I stand behind the usefulness of the legal approach, though, along with all feminists, I recognize that legal remedies, if they are remedies at all, are only one wing of an attack on oppression.

I suggest legal reform for a number of reasons. First, I am heartened by the jurisprudence of equality emerging within Canada, as noted above. Although neither fully secure nor consistent,[5] the fact that the Supreme Court of Canada has countenanced and cited feminist equality arguments is encouraging.[6] Second, with Cornell, I recognize that the symbolic realm, which inspires our imagination and into which we project our self-images for recognition, is significantly constructed and controlled already by law. Law can effectively pronounce individuals legal persons or not, can declare that groupings are families or not, can endorse lifestyles and sexual orientations or not, can define maternity, mental health, work, race, and marital status. As Cornell says, "Why enter the preservational economy of law at all? My answer is that we have inevitably already been entered into it. Our demand is to enter it differently, on the basis of the equivalent evaluation of our sexual difference" (Cornell 1995, 235).

Conclusion

In this chapter, I have argued that the practices of contract pregnancy and sex selection contribute to women's inequality, not primarily by immediate harms, but by the way they represent women. Endorsing contract pregnancy perpetuates the symbolic relegation of women to reproductive labor, as well as severely compromising the autonomy of the contracting pregnant woman. This compromise results from both the contractual pressures that apparently cannot be removed and the lack of adequate models for female self-constitution. Endorsing sex selection (within patriarchy) contributes to women's inequality by seeing the "threat" of a female child as sufficient reason to fund sex-determination clinics, to select male preembryos within the IVF procedure, and to offer ultrasound for fetal sex determination. This palpable illustration of the devaluing of the female is an impediment to the self-respect necessary to becoming a person and dominates the imaginary space, obstructing the "restoration of the feminine within sexual difference."

What legal remedies do I propose? Along with other Canadian feminists,[7] I endorse targeting not women but third parties to contract pregnancies, as well as refusing to enforce pregnancy contracts. Targeting third parties provides a disincentive to those who see a market opportunity but have little concern for the status or well-being of women and children.

Legal remedies to sex selection vary according to the technology in question. I would advocate a ban on preconception sex selection services, withholding funding and/or licensing from infertility clinics that offer sex choice within the IVF process, and requiring hospitals and clinics offering ultrasound to adopt a policy of nondisclosure of sex information. Such a policy, and the reasons for it, should be public knowledge so as to protect informed consent. An exception should be made where sex information has important medical relevance.[8]

These proposals will not receive unanimous support from feminist writers. Each has been hotly debated, and I admit to a good deal of ambivalence myself over both legal remedies in general, and these specific provisions in particular. My primary goal in this essay is less to defend the particular legal remedies sketched as to draw attention to what I consider to be an important line of reasoning about the threat to women's equality inherent in the practices of contract pregnancy and sex selection. If I only attract critical challenge and further discussion of this line of reasoning, I will nonetheless be pleased.

Notes

I wish to express grateful thanks to Jennifer Parks, whose research assistance and collaboration contributed enormously to this essay.

1. See, for example, Young 1990.

2. Cornell's book is a rich and complex argument, to which I cannot do justice in this brief exposition. I hope I represent it fairly enough to inspire those who haven't yet read it to do so.

3. Legal positivists might prefer the phrase "entitled to," although Cornell's usage reflects her commitment to a Kantian legal philosophy.

4. I recognize this claim is somewhat speculative, as, indeed, are all empirical claims about the long-term effects of practices on the public imaginary. On this, Arneson and I may just have to agree to differ!

5. See Lepofsky 1992.

6. See Majury 1991 and Mahoney 1992.

7. See the Quebec Counsel for the Status of Women 1989; and Overall 1993.

8. For an interesting defense of the morality of preferring to bear a nondiseased child, see Purdy 1995.

References

Annas, George. 1986. "Pregnant Women as Fetal Containers." *Hastings Center Report* 16 (6): 13–14.

Arneson, Richard. 1992. "Commodification and Commercial Surrogacy." *Philosophy and Public Affairs* 21 (2): 132–164.

Boetzkes, Elisabeth. 1994. "Sex Selection and the Charter." *Canadian Journal of Law and Jurisprudence* 7 (1): 173–192.

Cornell, Drucilla. 1995. *The Imaginary Domain*. New York and London: Routledge.

Feinberg, Joel. 1984–1988. *The Moral Limits of the Criminal Law*. New York: Oxford University Press.

Fudge, Judy. 1990. "What Do We Mean by Law and Social Transformation?" *Canadian Journal of Law and Society* 5: 47–69.

Hart, H. L. A. 1963. *Law, Liberty and Morality*. Stanford, CA: Stanford University Press.

Held, Virginia. 1993. *Feminist Morality: Transforming Culture, Society, and Politics*. University of Chicago Press.

Langton, Rae. 1993. "Speech Acts and Unspeakable Acts," *Philosophy and Public Affairs* 22 (4): 293–330.

Lepofsky, M. David. 1992. "The Canadian Judicial Approach to Equality Rights: Freedom Ride or Rollercoaster?" *National Journal of Constitutional Law* 1: 315.

MacIntyre, Sally. 1976. "Who Wants Babies? The Social Construction of 'Instincts.' " In *Sexual Diversions and Society: Process and Change*, D. L. Barker and S. Allen, eds. London: Tavistock.

Mackenzie, Catriona. 1992. "Abortion and Embodiment." *Australasian Journal of Philosophy* 70 (2): 136–155.

Mahoney, Katherine. 1992. "The Canadian Constitutional Approach to Freedom of Expression in Hate Propaganda and Pornography." *Law and Contemporary Problems* 55: 83.

Majury, Diana. 1991."Equality and Discrimination According to the Supreme Court of Canada." *Canadian Journal of Women and the Law* 4: 407.

McCormack, Thelma. 1989. "When Is Biology Destiny?" In *The Future of Human Reproduction*, Christine Overall, ed. Toronto: Women's Press.

Menon, Nivedita. 1993."Abortion and the Law: Questions for Feminism." *Canadian Journal of Women and the Law* 6: 103.

Meyers, Diana T. 1989. *Self, Society and Personal Choice*. New York: Columbia University Press.

Morgan, Kathryn Pauly. 1989. "Of Woman Born How Old-Fashioned!—New Reproductive Technologies and Women's Oppression." In *The Future of Human Reproduction*. Christine Overall, ed. Toronto: Women's Press, 60–79.

Overall, Christine. 1987. *Ethics and Human Reproduction: A Feminist Analysis*. Boston: Allen & Unwin.

———. 1993. *Human Reproduction: Principles, Practices, Policies*. Toronto: Oxford University Press.

Papp, Aruna. 1991. "A Matter of Gender." *Healthsharing* 12 (1): 12.

Purdy, Laura. 1995. "Loving Future People." In *Reproduction, Ethics, and the Law*. Joan C. Callahan, ed. Bloomington: Indiana University Press, 300–327.

Quebec Counsel for the Status of Women. 1989. *General Opinion of the Conseil du Status de la Femme in Regard to New Reproductive Technologies*, Government of Quebec.

Rothman, Barbara Katz. 1989. *Recreating Motherhood: Ideology and Technology in a Patriarchal Society*. New York: W. W. Norton.

R. v. Butler. 1992. Canada Law Books, Inc. *Supreme Court Reports* 1: 452.

R. v. Keegstra. 1990. Canada Law Books, Inc. *Supreme Court Reports* 3: 697.

Royal Commission on New Reproductive Technologies. 1993. *Proceed with Care*. Ottawa: Canada Communications Group Publishing.

Smart, Carol. 1989. *Feminism and the Power of Law*. London: Routledge.

Steinbacher, Roberta. 1981."Futuristic Implications of Sex Preselection. In *The Custom-Made Child? Women-Centred Perspectives*, Helen B. Holmes, Betty B. Hoskins, and Michael Gross, eds. Clifton, NJ: Humana Press.

Strickling, Bonnelle Lewis. 1988. "Self-Abnegation."In *Feminist Perspectives: Philosophical Essays on Method and Morals*, C. Overall, L. Code, and S. Mullett, eds., Toronto: University of Toronto Press; 190–201.

Thomson, J. J. 1971. "A Defense of Abortion." *Philosophy and Public Affairs* 1 (1): 47–66.

Tong, Rosemarie. 1995. "Feminist Perspectives and Gestational Motherhood: The Search for a Unified Legal Focus." In *Reproduction, Ethics, and the Law: Feminist Perspectives*, Joan C. Callahan, ed. Bloomington: Indiana University Press, 55–79.

Warren, Mary Ann. 1985. *Gendercide: The Implications of Sex Selection*. Totowa, NJ: Rowman & Allanheld.

———. 1994. "On the Moral and Legal Status of Abortion." In *The Problem of Abortion*, Joel Feinberg, ed. Belmont, CA: Wadsworth Publishing, 110.

Williamson, Nancy E. 1976. "Sex Preferences, Sex Control, and the Status of Women." *Signs* 1: 847–862.

Young, Iris Marion. 1990. *Throwing Like a Girl and Other Essays in Feminist Philosophy and Social Theory*. Bloomington: Indiana University Press.

8

Health Commodification and the Body Politic: The Example of Female Infertility in Modern China

Lisa Handwerker

Since the late 1980s, there has been a growth of infertility clinics and an increased interest in and use of new reproductive technologies (NRTs) in the People's Republic of China. In March 1988, television and newspaper reporters announced the birth of the first Chinese test-tube baby born to a thirty-nine-year-old peasant woman (Beijing Review, 1988). By December 1993, one of China's major teaching hospitals had produced over fifty-one test-tube babies through the combined efforts of in vitro fertilization (IVF) and gamete intrafallopian transfer (GIFT). While, at first glance, China's growing infertility industry seems inconsistent with a mandatory state birth reduction policy, it is consistent with China's emphasis on technological innovations in science and medicine as a key symbol of modernity (Lampton 1987; Simon and Goldman 1989; Suttmeier 1982; Zhao 1985). Furthermore, women, as the (re)producers of Chinese culture, figure prominently in the body politic.

As a result of these inconsistencies, tensions can be observed between those people in China who support reproductive technologies that promote births and those who do not. Despite these tensions, China, a country with an explicit policy to restrict births, has unwittingly encouraged the growth of an infertility industry and backed into a eugenic policy. This infertility industry has resulted from complex cultural processes and social structures that intersect with newly imported technologies. Rather than operate by coercive means, these reproductive technologies perpetuate a standard of reproductive normality—that is, the imperative for women to give birth to one child and the possibility to create a superior child for Chinese nation-state building. With the commodification of medicine, including the transfer of technologies from the West to China, one can observe the social and ethical implications of new technoscientific practices targeted for infertile women.

This essay, the first comprehensive ethnographic study of female infertility in modern China, draws on a number of anthropological data-collecting methods and sources including participant observation; semistructured and in-depth interviews with patients, doctors, other citizens, and representatives from international agencies; and textual analysis of popular materials such as magazine articles and patients' letters to doctors. I conducted my research in Beijing from 1990 to 1991 and during a brief follow-up visit in 1997.[1]

Relying on the example of China, a feminist analysis of the social and ethical implications of NRTs from a cross-cultural perspective can shed light on some questions we are grappling with at this historical juncture. Until recently, bioethics has ignored gender issues. This chapter examines reproductive technologies and specific cultural and social processes that inform them in Beijing, China. I explore the paradox of the "problem" of female infertility and the consequences of reproductive technologies within the context of China's population reduction policy. What accounts for the increased demand and use of infertility services, including IVF, in a country ostensibly committed to population control? How have recent health care reforms impacted treatment options for infertile women? What, if any, infertility treatments are covered by the Chinese government? Who benefits, with respect to gender, age, class, geographic location, and education, from the commodification of medicine, including the introduction of NRTs? What NRTs, if any, are appropriate to transfer to other countries with different political, social, and economic structures? In asking these questions, I hope to examine how infertile women's experiences both shape and are shaped by the body politic, including the birth planning policy and the commodification of medicine and medical care in modern China.

The Modern Politic: China's Birth Planning Policy and the Paradoxical Implications for Infertile Women

With census results indicating a rapidly growing population and Chinese leaders voicing concern about their increasing difficulty to meet national development plans (Wang 1988), the world's first mandatory one-child policy was instituted in China in 1979 (Croll, Davin, and Kane 1985; Greenhalgh 1990b). The government, anticipating noncompliance because of traditional valuing of sons for old-age insurance and lineage continuity, implemented both incentives and disincentives to encourage limiting family sizes. Briefly, those couples that agreed to follow the one-child policy were rewarded with job promotion, bonus money, better housing, and access to higher-quality child care and educational facilities. Conversely, couples that refused to comply faced punitive measures, including large fines, house demolitions, and job transfers (see, e.g., Banister 1987; Croll, Davin, and Kane 1985; Greenhalgh and Bongaarts 1985; Kane 1987; Potter and Potter 1990).

Despite the state's commitment to strictly control population, considerable variation in official attitudes and policy enforcement persists (see, e.g., Chow 1991; Greenhalgh 1990b; Ng 1990; Shue 1988). First, even national leaders are divided over the wisdom of the one-child policy (Zhang and Yang 1981, 29; Banister 1987, 371–376). Second, the Chinese regularly violate regulations, and bureaucrats busily make adjustments responding to conditions of local and global capital. Third, during my interviews with hundreds of women from several regions in China, it became clear that the term "one-child policy," as it is commonly known in the West, is a misnomer. In fact, in China it is referred to as the *jihua shengyu*, or "birth planning policy," which allows greater variability than the English version implies (Greenhalgh 1990a and 1990b; Potter and Potter 1990). Family planning officials have always found it easier to enforce the policy in urban areas through the *danwei* (work unit) than in rural agricultural areas, where the large extended family traditionally had both important symbolic and cultural meaning. State policing of the family has had a hand in restructuring the modern Chinese family on a more Western nuclear family model, as couples, and particularly women, come to be blamed for both overpopulation and infertility problems (Handwerker 1993).

The attempt to regulate population by the Chinese Communist Party (CCP) offers a lens through which to understand the construction of normative gender prescriptions. Certain forms of nationalism and state power require control over the body but are expressed differently for men and women (Foucault 1980; Parker, Russo, and Sommer 1992; Sawicki 1991). Enforcement of the birth planning policy has led to unprecedented surveillance and regulation of the Chinese social body in general and the female body in particular. While census statistics on men of childbearing age are rare, statistics on women's fertility and sexual and reproductive practices are gathered and widely circulated. Information is collected on the number of women of childbearing age, live births, abortions, miscarriages, pregnancies, and contraceptive use and knowledge. Furthermore, since the 1990s, semiannual gynecological exams have been required for some married women of reproductive age (Greenhalgh 1993, 32). Unmarried women are required to undergo premarital exams to detect any visible structural problems with their reproductive organs and/or family history of genetic abnormalities that might lead to the birth of a disabled child. When combined, this information becomes the means through which the party-state mobilizes social pressure to influence the reproductive and sexual practices of households/families (Anagnost 1988). In urban areas, mechanisms of female fertility surveillance were also achieved through charting menstrual cycles, granting permission for and documenting births, and distributing free contraceptives, especially intrauterine devices, which required implantation and removal by a medical professional. This birth campaign and its results exemplify what Michel Foucault (1980) has referred to as a "technology of normalization" that systematically scrutinizes, classifies, and controls anomalies of the body, especially the female body.

While the CCP's attempt to regulate female reproduction through birth planning has been, in many ways, an effective means of control and normalization, it paradoxically defines and reinforces infertile women as "other," or different from other women. One woman told me:

> I am sure the pressure to have children in China is greater than in any other country. I especially feel a lot of pressure from my work unit. I think if I didn't work the pressure might be less. . . . I feel so much pressure because of the mandatory birth certificate which provides me with permission to have a child. I have had to turn in my certificate the last three years because I couldn't have a child. I felt terrible. Here they give you permission and then you can't even give birth. I feel so humiliated to have to get a new certificate each year. This year I didn't even go to the planning authorities; they automatically gave me a new one in April.

This woman and many others felt discriminated against by the birth policy. Ironically, birth planning officials assume that all women of reproductive age should and will be fertile.

Within the context of population reduction rhetoric, one would anticipate that childless couples would be rewarded for their contribution to the state. Instead, childless women from urban areas see mothers with one child receiving such special state benefits as financial rewards, better housing, and a vacation on the Chinese holiday of "Children's Day." Peasant women are also under great pressure to have more, not fewer, children since the economic reforms of the 1980s led to the dismantling of rural collectives into individual plots (Anagnost 1989; Greenhalgh 1990a and 1990b; Potter and Potter 1990; Smith 1991). Today, larger families, especially sons, are considered an important factor increasing farming productivity. Overall, women from both rural and urban areas feel that pressure to have a child has increased in the 1990s. One woman I interviewed, summarizing the sentiment of many, said, "The one-child policy is really the you-*must*-have-one-child policy."

The Cultivation of a Superior Child: Eugenic Aspects of China's Birth Planning Policy

China's birth planning policy is a call not only for fewer births, but also for improved quality births. This has been implemented in two main ways: through the promotion of laws and through education. First, in their effort to improve China's overall population quality, leaders, including public health officials, proposed national legislation in 1993 to stop "abnormal" births through abortions and, if necessary, sterilizations (Tyler 1993). This proposed new legislation requires even tighter surveillance of pregnant women, for any woman diagnosed with an abnormal fetus would be advised to abort. This was not the first time such legislation had been proposed. Five years earlier, in the northwestern province of Gansu, legislation was passed to prevent any mentally disabled woman from having a child (ibid.).

Second, the birth planning institution plays an active role in promoting the development of the only child. The ideological education work of the one-child family policy has from its inception been accompanied by a series of informational materials on both eugenics and the correct care and education of the single child. Materials range from a practical discussion of infant care, such as bathing and feeding, to more complex educational concerns. In the early years of the one-child policy, the focus was on a child's physical well-being, especially in rural areas. Since then, the focus, particularly in urban areas, has turned to mental health and intelligence (Nathansen Milwertz 1997, 128).

The new ideal family includes one child who is physically and mentally perfect (ibid., 130). Responsibility for ensuring this lies with the parents, primarily the mother. To illustrate this point, in 1992 the First National Exhibition on Eugenics, Improved Childbearing, and Childrearing was held in Beijing (Nathansen Milwertz 1997, 130; Guofang 1992). The exhibition linked the one-child policy with the need for a superior only child. At one exhibition booth, doctors from the National Defense Scientific Working Committee Hospital No. 514 offered visitors the opportunity to test a computer program that calculates the optimal conception date to achieve a healthy and intelligent child. Calculations are based on information about the intellectual, emotional, and physical cycles of future parents. During my fieldwork, I spoke with several women who believed in the value of prenatal education. Over the course of their pregnancies, earphones were placed on their bellies for the fetuses to hear music. Interestingly, it is Western classical music that has been promoted as advantageous to fetal development and intelligence. Furthermore, mothers, after a birth, are also expected to (and often do) spend inordinate amounts of time, energy, and money raising their only child.

An important objective, if somewhat implicit, in the making of a superior only child is to ensure that the child is grateful to the parent for her/his development (Nathansen Milwertz 1997, 136). By establishing filial obligation, parents hope to ensure that the child will provide for them in their old age. This is particularly common now with a rapidly aging population and a government no longer willing or able to care for the elderly. In the case of infertile couples, the vacuum created by the collapse of institutions such as the extended family is now promised to be fulfilled by babies created through reproductive technologies such as IVF. The media promotes the belief that test-tube babies are also superior. One newspaper account describing one of China's first IVF babies, who was four and one-half years old at the time, reads:

> Meng Zhu has never been sick. She was trained to brush her teeth and put on her clothes at the age of three. She is taller than children her age by half a head. Her intelligence is also higher than children of her age. At present, Meng Zhu has mastered 500 Chinese words, can recite 20 poems, sing 30 songs and perform 10 different dances. She can also do simple arithmetic skillfully. (China Industrial and Commercial News, 27 Sept. 1993)

Overall, the cultivation of the superior or "perfect" only child has implications for both fertile and infertile couples that must be understood within the broader context of China's modernization plan, including higher living standards, access to more material goods, and better medical care.

A Burgeoning Infertility Industry

"I hope you aren't going to solve our infertility problem. China has too many people. We need less people, not more." Beijing taxi drivers, shop owners, sales personnel, students, friends, and colleagues repeated these admonishing words to me whenever I explained that I was conducting research on infertility. Today in China, *"renkou tai duo"* ("The population is too large") is the phrase used to describe a range of social deficiencies from long waiting lines, to inequitable health services, to high unemployment and inflation.

Not surprisingly, I learned that until 1985, the state's priority was to fund projects to curtail fertility rather than to assist infertile women. Before 1985, infertility research was lacking, and most hospitals did not even conduct the most basic infertility tests. One infertility specialist told me:

> Before 1985, most [biomedical] hospitals did not do much infertility investigation, even simple procedures like hydrotubation or hysterosalpingogram.[2] After the open door policy and WHO [World Health Organization] collaboration, large amounts of work was started on family planning. Although WHO suggested to start infertility [work], the other aspect of reproduction, not until 1985 did our government agree and they asked me to join the Infertility Steering Committee of WHO.

Since the party-state has been largely unsympathetic to infertility research and clinical medicine, state doctors addressing infertility issues must often justify their work and need for resources. One doctor I spoke with explained how much initial resistance she faced to her IVF research. Refusing to give her or her colleagues funding, the state acknowledged the importance of their work only after it was successful. Even her own son said, "What are you doing? Everyone is going to hate you." Other colleagues told her, "We already have a large enough population. Why treat infertile couples when they can help our population problem?" Another well-known endocrinologist specializing in infertility treatment refused to open an IVF clinic. Instead, she argued that in China, a country with limited goods, there needs to be more emphasis on the prevention of conditions leading to infertility, including chlamydia and other sexually transmitted diseases.

To defend the aggressive treatment of infertile women within the context of the population policy, some doctors raised humanitarian and scientific concerns. In response to a local reporter who asked why China should pursue test-tube babies when the hospitals are so busy with abortions due to the population explosion, the founder of China's first IVF clinic replied:

> China has a huge population. Surely, there should be family planning, which includes both contraception and abortion for control of birthrates and infertility treatment in some men and women. One may ask, "There are too many pregnant women, why treat infertility?" In this logic, since

China is overpopulated, why bother to treat diseases? Why help the disabled? Why cure the wounded and save the dying? Why set up hospitals? Why are we doctors needed?

Another physician argued that the effective implementation of the state family planning policy required that OB-GYN specialists be adept not only in contraception and birth control, but also in reverse sterilization in case of an accidental death of a child.

In an interview, a well-known physician stated that programs such as IVF and GIFT open up new avenues to promote basic medical research in genetics, immunology, and early embryology. Therefore, she argued, they are of major importance to the implementation of China's policy of family planning and eugenics and, thus, mark a nation's stage of scientific and technological development (Tu 1988, 303). The doctor further pointed out,

> I am sure one day [will] come when man [can] design his own body according to [society's] need, just as in making engineering designs. By [that] time, man will have been able to select chromosomes with the best genetic material for artificial fertilization so as to raise the qualities of the human body in all its aspects. Scientific search is endless and scientific thinking is beautiful. . . . Some scientists even dream of asexual human propagation. . . . All this is but a fantasy. But fantasies prove a thoroughfare to ideals. Isn't the birth of the test-tube baby proof of this? (Tu 1988)

Clearly, some practitioners are driven by the need to help infertile couples, especially women, who they feel suffer enormously in Chinese society. Others are motivated by their interest in science and in nation-state building. Still others are motivated by the potential for economic profits, a subject to which I now turn.

As mentioned at the outset of this chapter, in a newly competitive health economy, the treatment of infertility is a paradox. The paradox is that despite attempts to restrict births by the CCP, infertility is a potentially lucrative business for hospitals, doctors, and factories. One international population officer explained:

> We provide money annually towards the development of contraceptive research and technology in China. We presently fund five major factories for contraceptive production in various parts of China. During a recent site visit we discovered that officials have been using our money to secretly manufacture medicines to treat the infertile and sexually impotent because of the potential for large private profits. (Personal communication)

Baby making is also driven by the consumer, the infertile couple who desire a biological child and are especially vulnerable to market forces and public opinion. Infertile women cited six major reasons for their desire to have a child: (1) because "everyone wants a child"; (2) for lineage continuity; (3) to provide care in their old age; (4) to "relieve loneliness" and "add interest to life"; (5) societal pressure; and (6) family (spouse's or mother-in-law's) insistence. One woman cried, "If I don't have the ability to give birth, I am not a woman." Women expressed distinct pressures depending on their geographic location, social class, educational background, and work and family circumstances.

Childless couples searching for a solution to their infertility problems want the best care, and the medical profession teaches them that the best equals the most technical and most expensive care. The recent equation of Western medicine with modernity and prestige has led to the incorporation of biomedical techniques into an increasingly competitive market. Medical competence is now redefined by the quantity and variety of technology brought to bear on infertility. Whereas previously a woman could rely on free health care, this is no longer the case. In the past, free or low-cost health care was provided to state employees in the urban areas through the *danwei* and through the cooperative medical service in the countryside. From the 1960s until the early 1980s, peasants relied on a collectivized health care system. The communes had a medical care fund for farmers, which provided reimbursement for most medical expenses. The introduction of the contract responsibility system in the late 1970s contributed to the disintegration of the commune-based cooperative medical service.

During the 1980s, agricultural reforms made it possible for many Chinese living in rural areas to increase their incomes significantly, both by farming more productively and by entering into newly sanctioned and profitable sideline activities. Accompanying the new wealth was a rise in consumption, which in the health care domain resulted in a growing demand for more individualized and higher-quality services. Not surprisingly, among the outcomes was the relatively swift emergence of private providers. As early as 1984, more than 30 percent of all rural health care services were privately owned. One side effect of the economic reform was that peasants, for the first time since the 1950s, were allowed to travel freely into the cities to do their buying and selling, and sometimes even to live. Many of the newly wealthy peasants living in the suburban areas were able to access health care facilities in the cities; others were able to use their greater incomes to purchase additional health insurance coverage (Smith 1993, 764).

With the implementation of these reforms, health is increasingly being treated as a commodity, rather than a benefit provided by the state. The losers in such a situation are the rural poor, including both families of remote villages passed over by the reforms and families who have remained poor in relatively rich areas. In 1993, only 15 percent of China's villages are maintaining cooperative medical services (State to ask all to share, 1993). With the collapse of the rural health care system, patients from rural areas must now pay out of pocket for medical care. These recent changes hurt those least able to afford it. A survey by the Ministry of Public Health of 280,000 farmers in ten provinces and autonomous regions reported that 18 percent of those who were ill could not afford any medical treatment (Rural medical co-ops effective, 1993).

Dramatic changes in medical delivery can be witnessed in hospitals, which have been restructured by the modernization plans of the 1980s. Hospital staff are encouraged to work harder and more efficiently, with the added incentive of higher salaries (New hospital coverage, 1992). Since less money is available for hospital funding, the reforms have adverse economic consequences; hospitals plagued by financial problems now charge higher prices for what they say is "better treatment." Overall, changes in medicine require that families, both from rural and urban areas, share a larger share of medical costs in general and infertility costs in particular.

In one hospital I visited, women who were anxious to receive the best possible care from the "infertility specialist" purchased the slightly more expensive ticket (three yuan) to see her instead of the cheaper ticket to see a general practitioner (one yuan).[3] Since there was only one such specialist at this hospital, trained in the United States, all infertility cases were referred to her, regardless of the ticket price. This, I believe, is an indicator of the importance women place on their infertility problems and how much hope they have invested in the "miracles" of Western medical specialists and scientific technology. Thus, there are ample opportunities to exploit vulnerable patients.

Infertility treatment entails a wide range of services including both traditional Chinese and Western medicine. In the Western medicine clinic, expenses range from 150 yuan for a hysterosalpingogram to 2000–3,000 yuan for one IVF cycle (including an overnight hospital stay).[4] In the Chinese medicine clinic, costs are not necessarily cheaper. Contrary to popular beliefs, Chinese medicine can be very expensive. One peasant woman told me:

> I spent a total of 5000–6,000 yuan on Chinese medicine. I ate a *fu* daily which cost three yuan and in one year I spent a total of 900 yuan. Furthermore, this money all came from my family; my mother-in-law did not

contribute one yuan because she thought the treatment was useless. She wanted her son to divorce me. But I kept on eating the medicine. I ate so much medicine all those years and sometimes I felt so nauseous but still I kept on eating it. I believed if only I kept on taking the medicine I could become pregnant.

For poorer families with no medical coverage, seeking expensive treatment usually means pooling financial resources or borrowing money from relatives or neighbors. Many women, including the person quoted above, discussed their limited financial resources, which are needed for infertility treatment. Clearly, wealthier families are able to afford such new medical technologies as IVF while the poor cannot. However, one can no longer make assumptions that the person from the countryside is poorer and the person from the city is wealthier. The woman who gave birth to China's first test-tube baby lives in what was previously known as one of China's poorest regions. After the economic reforms, her family had accumulated and saved enough money over a few years to afford IVF treatment.

In urban areas, individuals who are state employees are theoretically still entitled to medical treatment and reimbursement from their work unit. I say "theoretically" because, while conducting fieldwork in Beijing, I learned about a document circulating among OB-GYN clinics that forbade the reimbursement of infertility treatment by the work unit. When I questioned two doctors at different hospitals about this document, they both agreed that it was subject to interpretation and, ultimately, not enforceable. One doctor defiantly said, "We get so many pieces of paper. We can ignore some. Sometimes papers even get lost or thrown in the garbage." My observations concur with her statement. Through participant observation and interviews in four OB-GYN clinics, I learned that reimbursement practices varied throughout hospitals across the country.

Many women from urban areas received full medical coverage for basic infertility treatment but not for NRTs.[5] But in order to receive 100 percent coverage, patients are required to seek treatment at a *hetongyiyuan*, or a designated hospital affiliated with their *danwei*. Unless prior approval is given, a woman who seeks care elsewhere may be denied reimbursement. While some patients seek reimbursement for each visit, I discovered that some women prefer to seek treatment at a hospital other than their *hetongyiyuan* to protect their confidentiality. In other words, a woman may not want others to know her business. A thirty-five-year-old female worker, married for seven years, told me:

In the beginning I wasn't willing to let others know about my infertility problem so I paid all the medical expenses by myself. In that way no one,

not even my colleagues, could find out. I have been to so many hospitals.
. . . For five years my pressure was so great. I believe it is a natural thing
for women to have babies. Since I can't, I feel this kind of psychological
pressure. But I now let other people know about my problem.

In other cases, women are required to cover some of their expenses.
Another thirty-five-year-old female, a state-employed worker diagnosed with endometriosis, confided:

> While I am *gongfei* [on public support], I do not submit bills to my local
> *danwei*. Rather, each month I receive twelve yuan for medical care. This is
> called *yiyaofei* [doctor and medicine support]. Each time I go to a doctor I
> must pay 20 percent of the medical expenses. For example, if this month
> I see a doctor and get medicine and the cost is fifty yuan I would pay 20
> percent or 10 yuan. At the end of the year my *danwei* will add up all the
> expenses that I paid and if that amount exceeds twelve yuan, they will
> reimburse me 100 percent. If the amount is less than twelve yuan I get to
> keep the money.

With the future of medical coverage for couples with infertility problems both expensive and uncertain, some women currently employed
by the state and entitled to 100 percent coverage are fearful that they
may lose the medical coverage they now have. This prompted at least
one infertile woman to seek a resolution to a long-term health problem.
This woman, a nurse in her late thirties, said:

> Since I was a teenager I have suffered from painful menstrual periods due
> to endometriosis. I am unable to conceive and after seeking all kinds of
> medical treatment over the years, I decided it's best to have a partial hys-
> terectomy. Although I am unable to have a child, I feel luckier than most
> other infertile women, because I know my husband loves me and would
> not divorce me. Only recently I gave up all hope to have a baby and do
> not want to suffer anymore from the endometriosis. My womb is useless
> (*wode zigong mei you yong*) and I don't want to have so much pain every
> month. Why keep on suffering when I already know I can't have a child?

When I asked her why she was intent to have a partial hysterectomy
now, especially after waiting so many years, lowering her voice, she
whispered to me:

> Lisa, China is not stable now (*Xianzai zhongguo bu wending*). It is not clear
> what will happen in the next few years and how much longer this govern-
> ment will remain in power. Now I have good *guanxi* through work and I
> can be assured I will receive good care. I can even pick a good doctor to
> operate on me, and receive 100 percent medical coverage for the opera-

tion. I will send the doctor a nice gift to ensure good relations. In the future, I have no such guarantee.

While the diversity of experiences among women seeking infertility treatments points to the difficulty of making generalizations in China, it seems that for the majority of infertile women, the commodification of health care, reflecting larger economic and political issues, has resulted in increased expenses and concerns about the future of the health care system in general and their own health in particular. In summary, an infertility industry has been unwittingly created by consumers and practitioners for a myriad of complex reasons including government mandates, market forces, social pressures, and people's expressed desires and needs.

Bioethical Considerations for Infertile Women Using IVF

Developments in reproductive medicine have made it possible to intervene in the procreation of children in unprecedented ways. Their full implications can be known only retrospectively. For example, I interviewed one woman who had undergone IVF, and her story sharply contrasted with the miracle account presented in the newspapers. She had suffered from infertility for ten years. After undergoing IVF, she "successfully" gave birth to a baby girl, but the consequences have been neither completely happy nor anticipated. Due to misunderstandings and disruptions in cultural processes, her husband has begun to beat her. Ironically, he never hit her before the birth.

Furthermore, patients are not always told or do not always understand the full and long-term medical consequences of reproductive technologies. IVF often leads to the use of other technologies or medical interventions. For example, a woman who becomes pregnant by IVF in China must deliver the baby by cesarean section. Babies born through assisted reproduction are considered more vulnerable and valuable than other babies, and, therefore, these women are not permitted to deliver vaginally. When IVF is successful, there is also the increased chance of multiple births. In China, doctors insert up to four embryos during IVF to increase the chance of achieving a pregnancy. Yet multiple children are unsupported by the Chinese government. This creates unnecessary economic and educational hardships and stigmatization of these children.

There are also serious questions about the appropriateness, distribution, and use of drugs in China. Many of the drugs necessary for IVF, such as Pergonal, are imported and expensive. This has created, at times, a demand greater than the supply. Questions arise about how to decide who should receive the limited supply. On one particular day in an infertility clinic, two patients had to draw straws to decide who would receive the drug. Clearly in China, NRTs and new forms of delivery and financing of health care services are emerging that have serious ethical implications that cannot be ignored for resource allocation and patient–practitioner relationships.

Conclusion

Until recently, the mainstream discipline of bioethics has ignored gender issues and feminist analyses within a broad social, political, and economic context (Wolf 1996, 4). A feminist bioethical perspective must necessarily address human reproduction, a field in which women have a lot at stake (ibid., 11). This chapter calls for more attention from a feminist bioethical perspective to the specific reproductive technologies and the specific cultural and economic processes that inform them. As feminist bioethicists explore the social and ethical implications of NRTs from a cross-cultural perspective, the example of China offers rich insights. It allows us to understand the complex ways in which technology has become not only a tool, but also an organizing principle, of human life with very specific implications for women's lives. In China, NRTs intersecting with complex cultural processes and social structures continue to perpetuate, rather than challenge, normative gender values equating womanhood with motherhood. In the 1990s, Chinese women are expected to produce one child in general and a mentally and physically superior child in particular.

The example of China enriches our own perspectives toward the use and practices of reproductive technologies in the United States. A cross-cultural perspective incorporating ethnographic data can help us identify the differences and similarities in values that govern the appropriation and use of NRTs in particular locales. Through the identification of specific cultural as well as biological processes, our notions of what is important in addressing infertility issues and transporting NRTs may be transformed. As we explore possibilities for creating future guidelines and taking appropriate social action, technology cannot provide sufficient direction. What is possible in the reproductive technologies arena may not be what all women need or desire. We need to think in more complex ways about women's bodies, our illnesses, and potential solutions. I hope to contribute to the ongoing discussion about the social and ethical implications of IVF by revealing the issues that shape its contemporary applications in China.

Notes

This essay is based on twelve months of research conducted in China in 1990 and generously supported by a joint grant from the Fulbright-Hays Doctoral Dissertation Award and the Committee on the Scholarly Communication with the People's Republic of China. Assistance was also provided by the Wanner-Gen Anthropological Association Dissertation Fund, a National Science Foundation Doctoral Dissertation Improvement Grant (No. BNS89-13347), and the Association for Women in Science. Additionally, the Soroptomist International Award and the University of California, San Francisco, provided dissertation writing support. Also, Stanford's Institute for Research on Women and Gender and UC Berkeley's Institute for the Study of Social Change provided me with important writing time. I appreciate the editorial insights of Anne Donchin and Laura Purdy. My work has benefited from many people, but for their ongoing support I thank J. Ablon, G. Becker, A. Clarke, T. Gold, the Handwerkers, A. Ong, and Y. Verdoner. Thanks also to Drs. Zhang Li Zhu, Zuo Wen Li, Chai, Gao, and Lu. I am warmly indebted to Dr. Yuan Hong and the women with whom I worked, for without them this project could not have been as comprehensive or as exciting. I also thank Meera Jaffrey, a dear friend, for my initiation to China. The above persons bear no responsibility for my data results or interpretations.

A version of this chapter will appear in *Medical Anthropology Quarterly*.

1. While single-locality studies are inadequate for drawing conclusions about all of China, I selected Beijing as my primary research site because it is a city rich in infertility clinics of both *xiyi* (Western) and *zhongyi* (traditional Chinese) medicine and the birthplace of the first test-tube baby. As the administrative capital from which family planning decisions are disseminated, Beijing is a lens through which to observe some tensions between policies discouraging and those unintentionally encouraging births.

2. Hysterosalpingogram is an X ray that requires the introduction of gas or dye into the uterus and tubes to check for blockage of the fallopian tubes.

3. In 1990, at the time of my research, five yuan was the equivalent of one U.S. dollar.

4. Within a fluctuating economy, it is difficult to state precisely how much this equals. For example, 2000–3,000 yuan or $400–500 might be equivalent to a farmer's one-month salary or a professor's one-year salary.

5. The general rule is that medical coverage does not include reimbursement for NRTs. But as in everything in China, there are always exceptions to a rule. I met one woman who believed she could even get reimbursed for IVF since she had personal connections with the health funds administrator.

References

Anagnost, Ann. 1988. Family violence and magical violence: Woman as victim in China's one-child family policy. *Women and Language* 11 (2): 16–22.

————. 1989. Transformations of gender in modern China. In *Gender and Anthropology*, Sandra Morgen, ed. Washington, DC: American Anthropological Association, 313–342.

Banister, Judith. 1987. *China's Changing Population*. Stanford, CA: Stanford University Press.

Beijing Review. 1988. First test-tube baby on mainland. March 21, 31 (12): 11 (Chinese Announcement appeared in the Renmin Ribao, March 1988).

China's first test-tube baby goes to school. 1993. *China Industrial and Commercial News*, September 27.

Chow, Rey. 1991. *Woman and Chinese Modernity: The Politics of Reading between the East and West*. Minnesota: University of Minnesota Press.

Croll, Elizabeth, Delia Davin, and Penny Kane, eds. 1985. *China's One-Child Family Policy*. London: Macmillan Press.

Foucault, Michel. 1979. *Discipline and Punish*. New York: Vintage.

————. 1980. *The History of Sexuality*, Vol 1: An Introduction. New York: Vintage.

Greenhalgh, Susan. 1990a. The evolution of the one-child policy in Shanxi, 1979–1988. *China Quarterly* 122: 191–229.

————. 1990b. The peasantization of population policy in Shanxi: Cadre mediation of the state-society conflict. Working Paper No. 21. New York: Population Council, 1–48.

————. 1993. The peasantization of the one-child policy in Shanxi negotiating birth control in China. In *Chinese Families in the Post-Mao Era*, Deborah Davis and Stevan Harrell, eds. Berkeley: University of California Press, 29–250.

Greenhalgh, Susan, and J. Bongaarts. 1985. An alternative to the one-child policy in China. *Population and Development Review* 11 (4): 585–617.

Guofang Kegongwei wuyisi yiyuan yiwusuo (National Defense Scientific Working Committee No 514 Clinic) 1992. Guogang kegongwei can zhan xiangmu jianjie [A Brief Introduction to the National Defense Scientific Working Committee Projects included in the Exhibition]. Guofang kegongwei jihua shengyu ligdao xiazu bangongshi banli. June 1992. Beijing.

Handwerker, Lisa. 1993. The hen that can't lay an egg (*Bu xia dan de mu ji*): The stigmatization of female infertility in late twentieth century People's Republic of China. Ph.D. diss. Joint Medical Anthropology Program, University of California, San Francisco and Berkeley.

Handwerker, Lisa. 1994. The hen that can't lay an egg (*Bu xia dan de mu ji*): Conceptions of female infertility in modern China. In *Deviant Bodies: Critical Perspectives On Difference In Science And Popular Culture*, Jennifer Terry and Jacqueline Urla, eds. Bloomington: Indiana University Press, 1994, 358–379.

Kane, Penny. 1987. *The Second Billion: Population and Family Planning in China*. London: Penguin Books.

Lampton, David M. 1987. *Policy Implementation in Post-Mao China*. Berkeley: University of California Press.

Nathansen Milwertz, Cecilia. 1997. *Accepting Population Control: Urban Chinese Women and the One-Child Family Policy*. Nordic Institute of Asian Studies Series. Surrey, England: Curzon.

New hospital coverage causes controversy. 1992. *China Daily,* December 9.

Ng, Vivienne W. 1990. *Madness in Late Imperial China: From Illness to Deviance.* Norman: University of Oklahoma Press.

Parker, Andrew, Mary Russo, Doris Summer, and Patricia Yaeger, eds. 1992. *Nationalisms and Sexualities.* New York: Routledge.

Potter, Sulamith Heins, and Jack M. Potter. 1990. *China's Peasants: The Anthropology of a Revolution.* Cambridge: Cambridge University Press.

Rural medical co-ops effective. 1993. *China Daily,* July 13.

Sawicki, Jana. 1991. *Disciplining Foucault: Feminism, Power, and the Body.* London: Routledge.

Shue, Vivienne. 1988. *The Reaches of the State: Sketches of the Chinese Body Politic.* Stanford, CA: Stanford University Press.

Simon, Dennis F., and Goldman Merle, eds. 1989. *Science and Technology in Post-Mao China.* Harvard Contemporary China Series, No. 5. Cambridge, MA: Harvard University Press.

Smith, Christopher. 1991. *China: People and Places in the Land of One Billion.* Boulder, CO: Westview Press.

———. 1993. (Over)eating success: The health consequences of capitalism in rural China. *Social Science and Medicine,* 37 (6): 761–770.

State to ask all to share cost of social insurance. 1993. *China Daily,* June 26.

Suttmeier, Richard P. 1982. *Science and Technology in China's Socialist Development.* Paper presented at the World Bank Science and Technology Unit, Project Advisory Staff, January.

Tu, Ni. 1988. Revealing the secrets of human reproduction: Recording the birth of a test-tube baby. (translated from unidentified magazine article).

Tyler, Patrick E. 1993. China weighs using sterilization and abortions to stop "abnormal" births. *New York Times,* December 22.

Wang, Feng. 1988. Historical demography in China. (Hawaii: East-West Population Institute) *Review and Perspective* 236: 53–69.

Wolf, Susan M., ed. 1996. *Feminism & Bioethics: Beyond Reproduction.* New York: Oxford University Press.

Zhang, X., and C. Yang. 1981. *Qianjin zhing de xin wenti* (New problems in the forward march). *Guangming Ribao* (Bright Daily, Beijing) 2 (September 29).

Zhao, Z. 1985. *Gaige keji tizhi, tuidong keji he jingsi, shehui xietao* (Reform the science and technology system, and promote its coordination with the economy and society). *People's Daily* 12 (March): 1–3.

9

Feminism and Elective Fetal Reduction

Mary V. Rorty

In the summer of 1996, the world press carried the news that a woman in England had requested reduction of one of her two naturally conceived, apparently healthy twins. Amid public outcry, offers to adopt the unwanted member of the pair if she would carry both to term, and donations of money to assist her if she raised both children, it was revealed that the requested reduction had, in fact, already been performed (Wallace 1996).[1] The physician involved commented, "Killing one healthy twin sounds unethical . . . [but it was better to follow the mother's wishes and] leave one alive than to lose two babies" (ibid., 20).

Is it unethical to abort one of two healthy twins at the mother's request? Does elective fetal reduction present problems for feminists concerned about bioethical issues?

Even as bioethics was beginning in the United States in the 1970s, feminists had begun addressing ethical issues in medicine. The women's health movement, including the Boston Women's Health Book Collective, was already having an impact upon women's encounters with established medicine. In 1979, funded by an NSF grant, a group of feminist biologists, physicians, philosophers, and sociologists met at Hampshire College to examine issues raised by reproductive technologies. The two volumes from the proceedings of that meeting were among the first collections of their kind (Holmes et al. 1980 and 1981). Feminists from different disciplines approach health care from different matrices, yielding differing perspectives on the implications for women and society. Indeed, the variety of disciplinary backgrounds

159

that contribute to feminist bioethics is one of the great strengths that feminism brings to contemporary bioethical deliberation.

The different feminist approaches to bioethical issues of common interest supplement, complement, and occasionally contradict one another. This chapter considers one recent and increasingly popular technology: fetal reduction. This procedure currently has two primary contexts of application: (a) as a "lifeboat" intervention, reducing the number of surviving embryos in an established pregnancy so as to increase the chances of successful delivery; and (b) as a "selective" or eugenic procedure to prevent the live birth of an anomalous fetus in a multiple pregnancy. I will consider arguments for and against extension into a third context: (c) as an elective procedure to allow expanded reproductive decision making to prevent natural multiple birth of normal twins.

Fetal Reduction

In 1978, Louise Brown, the first in vitro fertilization (IVF) baby, was born in England. Dr. R. G. Edwards and Dr. Patrick Steptoe returned to the mother's uterus an ovum fertilized in vitro that successfully implanted and was carried to term, thus opening a new era in assisted reproduction. The same year, with less public furor, a second medical innovation was heralded in England: Dr. Anders Aberg, by cardiac puncture, selectively aborted one of two fetuses in a twin pregnancy, performing the first fetal reduction (Aberg et al. 1978).

Despite different research trajectories, the two innovations were mutually reinforcing. For some assisted reproduction techniques may be too successful. Fertility treatments can stimulate the ripening of several ova simultaneously, leading to multifetal pregnancies. IVF procedures, in order to assure successful outcome of at least one live birth, may return several fertilized ova to the uterus for implantation, typically, three or four, although higher numbers have been recorded. Some may fail to implant; but if all survive transfer, premature delivery is almost certain. As fertility treatments have grown in popularity and success, multiple births have increased significantly worldwide.[2] But more is not always better. One of the ironies of assisted reproduction is a version of the law of diminishing returns: the more fetal sacs involved in a pregnancy, the lower the chance for successful live birth and the greater the chances of premature delivery, with its sequelae of serious physical and developmental injuries.[3]

The solution to Dr. Edwards's problem was found in Dr. Aberg's procedure. Intervention in an ongoing pregnancy to lower the number of fetuses carried to term has become a widely accepted procedure, termed "fetal reduction."[4] The procedure introduced by Aberg is currently applied in three contexts that are distinguished in the literature by different terminology.

Multifetal Pregnancy Reduction

By far the majority of the procedures first introduced by Aberg are interventions in fertility-enhanced multiple pregnancies: reductions for sheer fetal number. Reductions of naturally occurring pregnancies are rare. They are usually considered essential for successful live birth in multiples above four, and they are typically strongly recommended in triplet and quadruplet pregnancies as well. Although the standard varies across centers, multiple pregnancies are typically reduced to twins. Selection of which fetus to reduce is determined by physical accessibility and location in the uterus, rather than by any characteristics of the fetus.

Selective Reduction

The original procedure and some modifications of it are chosen for a different reason: to interrupt the development to term of an anomalous fetus. It is now possible to diagnose with relative reliability over two hundred handicapping disorders. Abortions when there is a "substantial chance that a child will be born with grave mental or physical disability" have been sanctioned by the medical profession since the 1960s, and "the medical literature betrays a widespread, usually unstated assumption by the profession that fetal life should be terminated whenever a serious congenital abnormality is strongly suspected" (Beck 1990, 181). Selective reduction, like its companion procedure, selective abortion of singleton pregnancies, prevents the birth of a severely damaged infant whose life, some argue, represents an increased amount of suffering in the world.[5]

Selective abortion is a different kind of choice than deliberate pregnancy interruption. The latter follows a decision not to reproduce,[6] while the former follows a decision not to carry this specific fetus to term because of its distinctive characteristics. The conceptual distinction is reflected in a legal one: in sixty-seven international jurisdictions reviewed by Rebecca Cook in 1987, only thirteen have legal provision for abortion upon request for various conditions, but forty-eight have provision for abortion for "fetal health," that is, "some degree of likely physical or mental impairment of a child who may be born on natural completion of pregnancy" (1989, 68).[7] In those few jurisdictions that allow abortion upon request of the mother, selective abortion may fairly be seen as one possible subset of elective abortion; but in other jurisdictions, a pregnant woman's own preferences may be subordinated to other considerations.

Elective Reduction

Recent cases (such as that cited at the start of this chapter) have raised questions about extending the procedure to include a woman's preference for just one live birth. This extension, already acknowledged de facto by some practitioners, would allow reduction as an elective procedure.

Extension of fetal reduction procedures as elective is currently problematic, from the standpoint of both the general public and some medical authorities. Professional reactions to elective abortion, like public ones, are ambivalent. Physicians, including many involved in fetal reduction procedures, may be willing to perform abortions or reductions "for medical reasons," including fetal anomalies, but reluctant to perform elective abortions, or "abortion on demand." One of the earliest practitioners of multifetal pregnancy reduction has declared himself unwilling to reduce below twins, even on parental request, lest physicians "become technicians to our patients' desires," tacitly equating nonmedical with "trivial" reasons.[8] Other centers accept parental preference as a sufficient reason for choice of number in the absence of contraindications when higher multiple pregnancies are being reduced.[9]

Elective fetal reduction on a naturally occurring twin pregnancy is anomalous in medical practice for several reasons. First, reduction for "sheer fetal number" usually arises in the context of higher multiples, not twins. Typically, a multiple pregnancy is reduced to two embryonic sacs; so a twin pregnancy looks much more like the result of a successful reduction than the indication for one. Second, there is considerable danger of losing the entire pregnancy following reduction. As the time lapse between intervention and pregnancy loss increases, the possibility of attributing the loss specifically to the intervention decreases, but all centers acknowledge the possibility. Recent summaries suggest that even in very experienced centers, the probability of loss of the entire pregnancy before twenty-four weeks ranges from 8 to 12 percent— higher than the rate of pregnancy loss for twins without any intervention.[10] Third, even if the intervention does not cause pregnancy loss, there is an increased chance in multiple births of one of the fetuses disappearing in the course of the pregnancy. Leaving two fetuses, rather than one, provides a "hedge" against the loss of the entire pregnancy.[11]

As the outcomes of fetal reductions improve, professional consensus may gradually swing toward a liberalization of attitudes toward elective reductions. Influential members of the medical community support this extension. In a recent article arguing for extension of reductions to allow elective terminations, Frank Chervenak, Laurence McCullough, and R. Wapner (1995) note that the status of "patient" is conferred on a fetus by a woman who has decided to carry it to term. Chervenak and McCullough summarize their position, which was expressed in more detail in their 1994 book, as follows: "the fetus is a patient only in pregnancies being taken to term. The moral status of the previable fetus as a patient depends upon the pregnant woman's decision to confer such status."[12]

In abortion decisions, the pregnant woman withholds the status of "future child" to the fetus; and in fetal reductions, to one or more of the fetuses, "something she is free to do as a matter of exercising her autonomy to set her own goals for her pregnancy" (Chervenak and McCullough 1994). The authors go on to suggest that the extension of fetal reduction to an elective procedure is "consistent with existing public policy in the United States and Great Britain" (Chervenak et al. 1995, 535).[13]

Feminist Pluralism and Fetal Reduction

Some feminist writers focus on the social implications of institutions or practices; others address the specific situation of individuals making decisions about their own lives. The two tendencies of feminism, to contextualize and to particularize, ideally reinforce rather than contradict each other. It is only when particular women understand the structural incentives that construct (and sometimes artificially constrict) their options that they are able to make rational and informed decisions about their lives. The particularizing moment in feminist writing on reproductive issues often focuses on individual autonomy and informed consent; the contextualizing moment often emphasizes the network of unequal power relations within which individual decisions are made. Feminists who battle for the freedom for individual women to make decisions affecting their lives must also address the social conditions that structure the alternatives open to them.

As we consider the question of the advisability of making fetal reduction an elective procedure, both the impact upon individual lives and the wider institutional and social context are important. Elective fetal reduction is analogous in some ways, and disanalogous in other ways, to already accepted medical practices. That something has been institutionalized as standard practice does not answer all the ethical questions about it; and our best ethical thinking is proactive, not retrospective.

Elective Abortion and Elective Reduction

Feminists largely agree about the importance of preserving legal and safe abortion to prevent unwanted births. Women who choose not to carry a pregnancy to term have the moral right—and in the United States, since *Roe v. Wade,* the legal right—to interrupt an established pregnancy by elective abortion. Feminists ask what justification there is for not extending the option of elective abortion to the new techniques. Christine Overall suggests that where the option of elective abortion is available, it sets a precedent that cannot consistently be denied to those who use new reproductive technologies:

> If women are entitled to choose to end their pregnancies altogether, then they are also entitled to choose how many fetuses and of what sort they will carry. Legal or medical policy cannot consistently say "You may choose whether to be pregnant . . . but you may not choose how many shall occupy your uterus." (Overall 1994, 155)

Overall is arguing for the appropriateness of selective reduction of anomalous fetuses in multiple pregnancies. But one could argue as well that the right to abort one fetus in a multiple pregnancy in a society where elective abortion is legal cannot be restricted to only handicapped fetuses. Since genetic anomalies are only one of several indications for elective abortions in current practice, many arguments that apply to selective reduction apply to elective reduction as well, at least within the first trimester.

Some feminists object to the restriction to the first trimester. The technique used for this procedure—cardiac puncture with potassium chloride (KCl) injection—can also be used to assure a stillbirth in cases of late abortion. The procedure involves a transabdominal needle insertion of potassium chloride into the fetal heart (Evans et al. 1998). In a recent exchange between Mark Evans and John Fletcher and feminist bioethicist Joan Callahan, Callahan argued that such a procedure, under some circumstances, should be considered acceptable, not only for selective abortion in the case of fetal anomalies, but also for late-term elective abortions even when no fetal anomalies are involved:

> I want to suggest that a critical question in deciding on the justifiability of using KCl injection in any late abortion is this: If this pregnancy yields an

infant and that infant survives, what are its life prospects? [I]f there is
good reason to believe that it would not be in some particular infant's best
interest to emerge as a person, either because of profound anomalies or
because there is no one standing by who will see to that infant's flourishing, it would be wrong to bring that infant forward.[14] (Callahan 1995,
278–9)

Women's Choice and Human Experimentation

Feminist concern about conditions constraining choice is a subset of a
broader bioethical emphasis on informed consent. If a woman's only
or main source of information about a procedure, such as fetal reduction, is from a practitioner or medical center with a professional interest in improving its own experience or techniques, and if the novelty
or incidence of the procedure is such that long-term outcomes are not
yet well known, there may be reason for feminists to worry about
whether minimum conditions can be met for informed consent to fetal
reduction.

What considerations govern individual choice in medical decision
making? The decision maker needs sufficient information to determine
effects on health outcomes—for herself and, in the case of pregnancy
reduction, for the surviving fetus. In light of the risks, she must compare benefits and harms and consider comparable risks, benefits, and
harms of alternative technologies or other routes to the same outcome.
The results of this information-gathering process are needed to ensure
rational decision making. Women who consider fetal reduction need
complete and realistic descriptions of the risk of the procedure, its effectiveness or efficiency (both in general and when performed by the
specific physician and medical center), and the possible side effects to
themselves and surviving fetuses. An important consideration is the
cost of the procedure, which is unlikely to be reimbursed by medical
insurance.

But sufficient information for rational decision making may be hard
to find. Because most of the literature on fetal reduction procedures
concerns multifetal pregnancy reductions, there is not much information on results in smaller multiples. For a procedure that may have
been performed fewer than two thousand worldwide, statistically reliable information about outcomes may not be available.

This problem is particularly troubling in the area of reproductive
medicine, where silence or deception about success rates, in IVF procedures particularly, has been widespread. Reports in the medical literature on success rates may measure success by different markers than

those important to patients. A successful reduction, for instance, may be followed after a period of time by loss of the pregnancy. One interesting ambiguity involves doubts about whether reduction eliminates the prematurity and reduced fetal size associated with multiple births. It is important for decision makers to know that low birth weights are still a feature of reduced pregnancies, as low birth weight may be one of the factors contributing to health and development factors.[15]

There is a long distance between experimental procedures and standard medical practice. The procedures associated with fetal reduction, because of their relative novelty, intricacy, and infrequency, are much closer to experimental than to standard. One of the most problematic aspects of new reproductive technologies and techniques is that their success can be determined only by implementing them on women. Thus reproduction is one of the areas where the line between research and human experimentation is hardest to draw.[16] It is no wonder that practitioners and researchers struggle to restrict the procedures to cases where the results of intervening are likely to be better than the results of not intervening. The relative infrequency of elective reductions on the list of procedures summarized by various centers is as readily explained by this hypothesis as by any other. Feminists seeking to make new techniques and technologies more widely available are well advised to consider the warnings by feminists concerned to protect women from insufficiently developed experimental procedures.

An extremely worrying aspect of the current practice standard is the fact that, since twin reductions are not offered as an elective option, and thus not discussed in advance, some of the twin reductions explicitly discussed in the literature are performed under conditions that give every reason for concern to bioethicists worried about conditions for consent. In one of the few nonselective twin reductions discussed by one practitioner, a desperate woman pleading for alternatives to the termination of her entire twin pregnancy in the face of straitened economic circumstances is apparently told on the operating table of the possibility of reduction as an alternative (Chervenak et al. 1995). One must wonder whether the scenario described allows opportunity for informed deliberation. It is a cause for concern as well that the option of elective reduction is presented rhetorically as a response to desperate women's demands (Raymond 1994).

Elective Reduction and Selective Abortion

Thanks to advances in medical technology and prenatal diagnosis, we can learn earlier and earlier, more and more, about the once-hidden contents of the pregnant womb. For instance, the same time as the woman in the above case learned that she had two fetuses, she could have learned as well, by chorion villus sampling, whether there were any karyotypic anomalies of each fetus, and possibly a great deal more—whatever can currently be determined by genetic analysis, from the pathological to the trivial.

In elective reduction, how will it be decided which fetus will be carried to term and which not? Any information about either fetus that can distinguish them, in a context where only one will survive, discriminates between them to the disadvantage of one. Any characteristic can take on disproportionate significance. Elective reduction, where only one can survive, becomes de facto selective reduction. Is this an important conflation for feminists?

Selective abortion has received much less discussion in the feminist literature than elective abortion. What discussion there is often occurs in the context of specific instances of selective abortion, rather than addressing the practice in general. Sex selection is troubling to some feminists; and some disability activists have expressed distress about cultural attitudes toward people with disabilities that are manifested in some selective abortions for genetic anomalies. The medical profession has, to a considerable extent, welcomed the advent of prenatal genetic testing as a technology that increases the range of options available for prebirth planning, although the issue of control remains an important one: Who should decide whether a characteristic is of sufficient weight to justify a decision to terminate a pregnancy?

One concern is that, if some way of choosing between potentially viable fetuses must be found, one of the most accessible pieces of information, in the absence of any disabilities in either, is the sex of the fetuses. Is elective reduction another avenue for sex-selective abortion? There seems little ground for concern on this front in most Western societies, where family balancing and preference in birth order are more frequent concerns than clear preference for children of one gender over the other. Abortion practices in other societies, however, where the sex ratio of aborted fetuses tends to suggest male children are preferred to female children, are more problematic. Such systematic preference of male over female children manifests a categorical devaluation of the female and provides a reason for feminists to disapprove of sex-selection abortion. On the other hand, the principle articulated by Overall of "allowing a woman to choose how many and what kind" of fetus to carry to term must, if exercised consistently, allow this decision. Further, the suggestion by Callahan that considerations of minimal or reduced "life chances" are a sufficient reason for abortion, even late-term, clearly justifies the abortion preference of individual women in cultures in which gender constitutes a severe life disadvantage. Feminists can deplore the social conditions that constrain individual women to such choices and can work toward changing those conditions; but at the same time, it seems unfair to force individual women and their children to bear the whole social cost of such changes by restricting their otherwise available options.[17]

Abortion for "fetal indications" is an option offered to pregnant women in standard obstetrical practice. Information about possible disabling conditions for prospective children can be welcomed by individuals who seek more information and more control over their reproductive decisions. But some feminists and other activists have expressed concern about selective abortion for genetic reasons—either as a social policy or as one subset of individual decisions.

> Since the advent of prenatal testing and selective abortion, there have been feminist critics who have feared that the detection and elimination of "imperfect" fetuses will become socially if not legally mandatory. . . . Some predict that parents will come to demand more and more "perfect" children, and will thus abort for more and more minor (or merely imagined) fetal defects.[18] (Warren 1994, 117)

The abortion of fetuses with abnormalities such as Down's syndrome or spina bifida could lead to a loss of support and concern for persons born with similar disabilities.

> For example, there is a special moral significance to the termination of a fetus with a disability such as Down syndrome. The use of prenatal diagnosis followed by abortion or selective termination may have eugenic overtones when the presupposition is that we can ensure only high quality babies are born and that "defective" fetuses can be eliminated before birth. The fetus is treated as a product for which "quality control" measures are appropriate. (Overall 1994, 156)

The practice of selective termination may have a tendency to promote overall negative attitudes about disability in several ways: by suggesting that parents who bear children with disabilities are somehow to blame for doing so, or by creating a subset of language about genes ("bad" genes or "defective" genes) that may be transferred to the individuals who bear those genes (Kaplan 1994).

The concerns of disability activists about the social impact of the practice of selective abortion for predictable anomalies merit serious consideration. But feminists tend to be wary of proscribing the behavior or choices of other women, and they are unlikely to suggest lists of cans and cannots for selective abortion. Since one of the reasons most frequently given for selective termination is the probability of suffering to the resulting child if the pregnancy is brought to term, Down's syndrome is an indication for termination that many find very troubling. Recent experience with people with Down's has altered our expectations about their length of life, educability, and life possibilities. On the other hand, this increasing attention to the well-being of actual children with Down's has developed in the same period of time in which prenatal diagnosis and selective abortion for Down's has become increasingly prevalent. This suggests that disability activists' fears that selective abortion will have a negative impact on actual people with the relevant disability seems, in this case at least, to be unjustified. Certainly any form of "disability paternalism" that would curtail all selective abortions would be resisted by feminists.

How realistic is the fear that reduction for twins will contribute to negative eugenic attitudes, conflating "elective" with "selective" (explicitly discriminatory) abortion? In some practices, scrupulous care is taken to minimize the information given about the fetuses if reduction is to be an option. The placement of the fetuses in the womb, rather than any particular characteristic of the fetus itself, determines which fetus will survive reduction. Punctilious attention to such "impersonal" determinants may minimize dystopian fears about frivolous and trivializing choices concerning the kind of child to bear; but fears of the "frivolous" mother who will select against an otherwise normal child for trivial aesthetic reasons must in any case be balanced against fears of the perfectionistic physician (or reluctant insurance company), who may be equally unable to distinguish morally relevant criteria.

Conclusion

All forms of reproduction take place within the bodies of women, and the development of new technologies and techniques in the area of reproduction involve women as subjects. Feminists have good reason to carefully scrutinize technological developments in this area. Feminist bioethics stands between the occasionally competing obligations of increasing women's control over their lives and advocating for their interests in the face of structural constraints that may be indifferent to the interests of women, either as individuals or as a class. Fetal reduction, an expanding medical innovation of the last few decades, represents a typical problem in an atypically vulnerable area: a risky procedure in the volatile practice of assisted reproduction.

Since the success of new procedures is a function of the continuing improvement of techniques and technology, there is a professional incentive to expand the range of indications for reduction procedures. In light of the parallel concerns of feminist bioethics with the decisions of individual women and the larger context within which those decisions occur, what contributions can feminists make to the continuing discussion of fetal reduction?

First, inform individual decision makers. The high degree of correlation between reduction procedures and fertility treatments suggests that information about reduction options needs to be thoroughly discussed before fertility treatments begin. Possibly perfected reduction procedures in the future might make elective twin reduction an appealing option for pregnant couples. But for the present, the role of feminist bioethics may be to concentrate on urging women who are considering it as an elective procedure to fully inform themselves of the risks, the available alternative options, and considerations for and against the procedure. They need to insist also that the discussion with practitioners be open enough to allow for adequate advance deliberation. There is some concern that fetal reductions, like other technologies associated with reproduction, contribute to the alienation of women from their own reproductive lives: "Medically risky interventions may do more to extend the power and wealth of the male dominated medical profession than to enable women to make authentic decisions about their own reproductive lives" (Warren 1994, 117).[19] But like many other technological and scientific innovations, reduction procedures bear as much hope for improvement of our lives as fear of worsening them.

Second, contextualize the issue. Fetal reduction is inseparable in practice from two larger issues: the fertility industry and the economics of child rearing. Race and class issues permeate the controversial area of fertility: birth for some is undesirable, a drain on public resources; birth for others is considered infinitely desirable, worth any expense and danger. Regulation and adequate information are hard to find. Social scrutiny of the fertility industry remains a high priority for feminists everywhere. Further, individual decisions about how many children to bear and under what circumstances are often driven by financial considerations. Social policy could offer coherent and consistent incentives concerning these sensitive issues, but it seldom does. Feminist scrutiny of public policies can be more fruitful than questioning the choices of individuals.

Notes

1. Note that Wallace's article in *Maclean's* (1996, 20–21) uses the term "selective termination"; the procedure referred to is most accurately termed "elective fetal reduction" and is called so herein.

2. Twins occur naturally about once in 90 pregnancies, triplets about once in 8,000 pregnancies, and quadruplets about once in 700,000 pregnancies. More recently, the number of multiple births of all orders has increased in several developed countries, an increase generally attributed to fertility treatments. "In England, for instance, there was a four-fold increase in quadruplets from 1946 to 1985; in Germany, comparing 1950 to 1980, there was a three-fold increase in triplets and a 23-fold increase in the incidence of quadruplets. In Japan and Belgium, the number of triplet births has also risen dramatically." See Frederiksen et al. 1992, 8; Papiernick, 1990.

3. The mean gestational age of twins, triplets, and quadruplets is typically thirty-seven, thirty-three to thirty-four, and thirty-one weeks, respectively. The relative risk for neonatal death in the case of twins is eight times that for singleton births; for triplets, fifteen times, and for quadruplets, twenty times the singleton birth risk. Among survivors, there is an increased incidence of disabilities largely related to prematurity (Frederiksen et al. 1992, 8).

4. The terminology applied to the procedure varied widely in earlier years; recently, a consensus about terminology has emerged. See Berkowitz et al. 1996; see also ACOG Committee on Ethics Opinion 1992. In the fifteen years in which the procedure has been practiced, several techniques have been developed. Cardiac puncture with KCl injection has emerged as the professional standard. Wherever appropriate, this procedure is used for all forms of fetal reduction. For a discussion of various techniques, see Golbus et al. 1988; esp. 341–342; Berkowitz et al. 1996.

5. This explanation for the introduction of selective abortion practices is commonly given; cf. the work of John Harris, especially his defense of selective abortion in *Wonderwoman and Superman* (1992). Some of the diagnosable genetic anomalies for which selective abortions or reductions are performed—physical anomalies or trisomies 13 and 18, or Hurler's syndrome, with which the first reduced fetus was diagnosed—can clearly be predicted to produce considerable suffering in those afflicted by them. Others, notably trisomy 21 (Down's syndrome) are more problematic; it is much harder to argue that children with Down's "suffer considerably," but Down's syndrome is one of the most frequent indications for selective abortion.

6. Feminist Christine Overall, for instance, distinguishes between the negative right not to reproduce, which she considers unconditional, and a positive right to reproduce, which, she suggests, might be conditional. See Overall 1987 and 1994.

7. Some jurisdictions specify that the risk must be extreme or represent a severe handicap or disability. In those jurisdictions that allow abortion upon the pregnant woman's request early in pregnancy, abortion at a later stage may still require justification based on fetal health or maternal health (Cook 1989, 68).

8. The phrase "technicians to our patients' desires" comes from an article by Mark Evans, author of many articles on multifetal pregnancy reduction (see, e.g., Evans 1990). Feminists are unlikely to find this line of argument persuasive, for it smacks of arbitrariness: "My reasons are medical; yours are triv-

ial." If we are to distinguish between valid and invalid reasons for wishing to terminate (or reduce) pregnancies, the reasons offered by the patients must be given the same serious consideration as those offered by the physicians.

9. "The question of which multifetal pregnancies to reduce and how many fetuses to leave is moot in some instances. The approach at Northwestern has been to allow the patient to participate in the decision-making, permitting the couple to choose whether one or two or three fetuses are left, while fully informing them of the potential risks and possible overall pregnancy outcome of their choice" (Frederiksen et al. 1992, 12).

10. In a survey of reductions, Evans and colleagues reported the loss of the entire gestation at up to twenty-four weeks of gestation as 16.2 percent in 1988–1991, dropping to 8.8 percent in 1991–1993, showing a "very steep dramatic learning curve" as the center surveyed gained more experience with the procedure (Evans et al. 1995).

11. Seoud et al. (1992) reported that of twenty-six patients in their study with triplet gestations, eleven patients (about 30 percent) spontaneously lost one or more (but not all) of the fetuses. The "vanishing fetus" cases, where in the course of a pregnancy one or more members of an initially multiple pregnancy disappear without any external intervention, have been confirmed by other researchers.

12. The qualification that a fetus is a patient only if that status is granted it by the woman who carries it, a qualification of earlier paternalistic stances often adopted by medical practitioners, is one that is very sympathetic to feminist positions and is expanded at greater length in McCullough and Chervenak 1994.

13. The legality of selective reduction of pregnancy in England is contested in a recent article by Alison Hall, who calls attention to the legal terminology of "miscarriage" and of pregnancy termination; she raises the question of whether British law as currently written does in fact allow selective reduction, since the "reduced" fetus is not expelled from the womb and the pregnancy is not terminated (Hall 1996).

14. Callahan's suggestion that late-term abortion for nonanomalous fetuses can be justified solely in terms of life chances for the infant if born, like many issues surrounding the abortion debate, is not a matter of universal agreement among feminists. Late-term abortion in general is among the more highly contested forms of abortion in contemporary medical practice, and most writers on the topic consider maternal health to be its primary justification. See Callahan's comments on Overall's response to Callahan 1995 in the introduction to her own book, *Reproduction, Ethics, and the Law*.

15. Some of the literature on fetal reduction reports that the retained fetuses nonetheless are smaller than typical twins in unreduced pregnancies and are hence liable to more health hazards than infants larger at birth. Cf. Alexander, Hammond, and Steinkampf 1995: in a study of eighty-two pregnancies—thirty-two reduced versus fifty nonreduced twin pregnancies—the mean length of gestation was shorter by two weeks in reduced pregnancies; the mean birth weight was lower by one-fifth. The authors conclude that "multife-

tal pregnancy reduction does not reverse completely the decreased gestational age and impaired fetal growth associated with high order multiple pregnancy," adding that "this effect is more pronounced when a larger number of fetuses is reduced" (ibid., 1203).

16. Feminist Janice Raymond despairs of the whole network of options made available by the new reproductive technologies: "Fetal reduction is another technical fix launched as a corrective to a past technical fix that has failed. Yet few see the absurdity of treating one disaster with another" (Raymond 1994, 131). From a different perspective, Raymond's description makes the whole prospect sound less appealing. Perhaps, as part of informed consent, a description such as Raymond's should be included in every information packet dealing with pregnancy reduction.

17. In an interesting analysis of feminist thinking on this issue, disability activist Anita Silvers (in this volume, chapter 10) argues that it is inconsistent for feminists to disapprove of sex-selective abortion and yet approve of selective abortion for other fetal characteristics. She attributes this inconsistency to an unacknowledged prejudice against people with disabilities.

18. This critique, put in the mouth of feminists here, sounds discouragingly similar to the concern of some physicians that women will choose to abort for "trivial" reasons (see note 8), a feminist "maternalism" to match physician "paternalism." Mary Anne Warren, like Overall, eventually concludes that the considerations raised suggest that women need to be more broadly informed but not forbidden access to new reproductive technologies: "I think it is important to listen to such warnings . . . nevertheless I remain convinced that women who have accurate and adequate information and affordable legal voluntary access to a wide range of reproductive technologies for the most part will make reasonably sound decisions about the use of those technologies—and that in any case they have the right to make those decisions themselves" (Warren 1994, 118).

19. As concurring opinions, Warren cites Arditti et al. 1984; Callahan and Callahan 1984; Stanworth 1987; Spallone 1989; Donchin 1989; and Warren 1988. See also Raymond 1994.

References

Aberg, A., F. Mitelman, M. Cantz, and J. Gehler. 1978. "Cardiac Puncture of Fetus with Hurler's Disease Avoiding Abortion of Unaffected Co-Twin." *Lancet* 190 (2): 190–191.

ACOG (American College of Obstetricians and Gynecologists) Committee on Ethics Opinion. 1992. "Multifetal Pregnancy Reduction and Selective Fetal Termination." *International Journal of Gynecology & Obstetrics* 38: 140–142.

Alexander, James, Karen Hammond, and Michael Steinkampf. 1995. "Multifetal Reduction of High Order Multiple Pregnancy: Comparison of Obstetrical Outcome with Nonreduced Twin Gestations." *Fertility and Sterility* 64 (6): 1201–1203.

Arditti, Rita, Renate Duelli Klein, and Shelley Minden. 1984. *Test Tube Women*. Boston: Pandora Press.

Beck, Malcolm N. 1990. "Eugenic Abortion: An Ethical Critique." *Canadian Medical Association Journal* 143 (3): 181–184.

Berkowitz, R. L., L. Lynch, J. Stone, and M. Alvarez. 1996. "The Current Status of Multifetal Pregnancy Reduction." *American Journal of Obstetrics & Gynecology* 174 (4): 265–272.

Callahan, Joan C. 1995. "Ensuring a Stillborn: The Ethics of Lethal Injection in Late Abortion." *Journal of Clinical Ethics* 6 (3): 254–263; also published in Joan C. Callahan, ed. *Reproduction, Ethics, and the Law* (Bloomington: Indiana University Press).

Callahan, Sydney, and Daniel Callahan, eds. 1984. *Abortion: Understanding Differences*. New York: Plenum Press.

Chervenak, Frank A., Laurence B. McCullough, and R. Wapner. 1995. "Three Ethically Justified Indications for Selective Termination in Multifetal Pregnancy: A Practical and Comprehensive Management Strategy." *Journal of Assisted Reproduction and Genetics* 12 (8): 531–536.

Cook, Rebecca J. 1989. "Abortion Laws and Policies: Challenges and Opportunities." *International Journal of Gynecology & Obstetrics*. Suppl. 3: 61–87.

Donchin, Anne. 1989. "The Growing Feminist Debate Over the New Reproductive Technologies." *Hypatia* 4 (3): 136–149.

Evans, Mark I., M. May, Arie Drugan, John C. Fletcher, Mark P. Johnson, and R. J. Sokol. 1990. "Selective Termination: Clinical Experience and Residual Risks." *American Journal of Obstetrics & Gynecology* 162: 1575.

Evans, Mark, Ralph L. Kramer, Yuval Yaron, Arie Drugan, and Mark P. Johnson. 1998. "What Are the Ethical and Technical Problems Associated with Multifetal Pregnancy Reduction?" *Clinical Obstetrics & Gynecology* 41 (1): 46–54.

Evans, M. I., M. Dommergues, M. P. Johnson, and Y. Dumez. 1995. "Multifetal Pregnancy Reduction and Selective Termination." *Obstetrics & Gynecology* 7 (2): 126–129.

Frederiksen, Marilyn C., L. Keith, and R. Sabbagha. 1992. "Fetal Reduction: Is This the Appropriate Answer to Multiple Gestations?" *International Journal of Fertility* 37 (1): 8–14.

Golbus, Mitchell S., Nona Cunningham, James D. Goldberg, Robert Anderson, Roy Filly, and Peter Callen. 1988. "Selective Termination of Multiple Gestations." *American Journal of Medical Genetics* 31: 339–348.

Hall, Alison. 1996. "Selective Reduction of Pregnancy: A Legal Analysis." *Journal of Medical Ethics* 22: 304–308.

Harris, John. 1992. *Wonderwoman and Superman: The Ethics of Human Biotechnology*. New York: Oxford University Press.

Holmes, Helen B., ed. 1994. *Issues in Reproductive Technology*. New York: New York University Press.

Holmes, Helen B., Betty B. Hoskins, and Michael Gross, eds. 1980. *Birth Control and Controlling Birth*. Clifton, NJ: Humana Press.

———. 1981. *The Custom-Made Child?* Clifton, NJ: Humana Press.

Kaplan, Deborah. 1994. "Prenatal Screening and Diagnosis: The Impact on Persons with Disabilities." In *Women and Prenatal Testing: Facing the Challenges of Genetic Technology*, Karen J. Rothenberg and Elizabeth Thomson, eds. Columbus: Ohio State University Press, 57–58.

McCullough, Laurence B., and Frank A. Chervenak. 1994. *Ethics in Obstetrics and Gynecology*. Oxford: Oxford University Press.

Overall, Christine. 1987. *Ethics and Human Reproduction: A Feminist Analysis*. Boston: Allen & Unwin.

———. 1994. "Selective Termination in Pregnancy and Women's Reproductive Autonomy." In *Issues in Reproductive Technology*, Helen B. Holmes, ed. New York: New York University Press, 145–160.

Papiernick, Emile. 1990. "Financing Multiple Births: A Personal Point of View." *International Journal of Fertility* 35 (6): 330–332.

Raymond, Janice G. 1994. *Women as Wombs*. San Francisco: HarperCollins.

Rothenberg, Karen J. and Elizabeth Thompson, eds. 1994. *Women and Prenatal Testing: Facing the Challenges of Genetic Technology*. Columbus: Ohio State University Press.

Seoud, M. A. F., J. P. Toner, C. Kruithoff, and S. J. Muasher. 1992. "Outcome of Twin, Triplet, and Quadruplet In Vitro Fertilization Pregnancies: The Norfolk Experience." *Fertility and Sterility* 57: 825–834.

Spallone, Patricia. 1989. *Beyond Conception: The New Politics of Reproduction*. Granby, MA: Bergin & Garvey Publishers.

Stanworth, M. 1987. *Reproductive Technologies: Gender, Motherhood, and Medicine*. Minneapolis: University of Minnesota Press.

Wallace, B. 1996. "When One Fetus Lives and One Dies: A British Doctor Sparks a Public Furor Over the Ethics of Selective Termination." *Maclean's Magazine* 109 (34): 20–21.

Warren, Mary Anne. 1988. "IVF and Women's Interests: An Analysis of Feminist Concerns." *Bioethics* 2 (1): 37–57.

———. 1994. "Abortion: New Complexities." In *Issues in Reproductive Technology*, Helen B. Holmes, ed. New York: New York University Press, 113–122.

10

On Not Iterating Women's Disability: A Crossover Perspective on Genetic Dilemmas

Anita Silvers

"When I was in a wheelchair, one man wouldn't let me in his store. You can't imagine how lucky we are when we are whole."

Sheila Lukins, cofounder of the Silver Palate food shop
and author of five cookbooks[1]

"Because women's situations are so various and the forms of oppression we suffer are usually multiple, most feminist issues are inextricably involved with questions of . . . justice and bias," Alison Jaggar observes in her discussion of feminist ethics.[2] The immediate purpose of this essay is to explore such an issue, one that arises when women find themselves identified as disabled. I have a longer-range objective as well: to bring feminism's strengths to bear against disability discrimination, a form of bias with a long and troubling history for women.

Women with physical, sensory, or cognitive impairments are increasingly insisting that they are marginalized in feminist theory to an extent little different from how patriarchal society devalues them. Thus, Virginia Kallianes and Phyllis Rubenfeld write: "Disabled women have begun to articulate their criticism of . . . the feminist movement's failure to integrate into its agenda the perspectives of disabled women."[3]

Fortunately, recent work in feminist ethics and biomedical ethics has begun to change the picture, as influential feminist scholars like Martha Minow, Susan Sherwin, Rosemarie Tong, and Iris Marion Young

177

have commented on the confluences of disability discrimination with gender oppression. And the cultural criticism of feminists Katherine Pyne Addelson, Susan Bordo, and Elizabeth Spelman, among others, illumines the situations of all women who suffer because their bodies or minds fall away from the culture's standards of perfection.

This chapter will combine insights from feminist and disability scholarship to show how a disability perspective would inflect feminist approaches to biomedical ethics. In addition, it will illustrate how such refined feminist insight can help resolve what other writers think of as genetic dilemmas precipitated by disability. Key here is a new model of disability that has recently emerged to inform U.S. law and public policy. This model, I argue, both fortifies and is strengthened by feminist theory.[4]

What is meant by "a disability perspective"? The conviction that disability is a social construction rather than a biological fact emerged in the 1970s from the way disability advocates saw their own situations and where they perceived the conditions that limited them to lie. Disability studies scholar Mike Oliver writes:

> the social model of disability has been the foundation upon which disabled people have chosen to organize themselves collectively. This has resulted in unparalleled success in changing the discourses around disability, in promoting disability as a civil rights issue and in developing schemes to give disabled people autonomy and control over their own lives.[5]

The social model of disability was further developed during the 1980s in the new field of disability studies. It attributes the dysfunctions of people with physical, sensory, and cognitive impairments to their being situated in environments that are built and organized in ways hostile to them rather than to their biological conditions. This reconceptualization of what it means to be disabled has become a tenet of U.S. public policy because it frames the 1990 Americans with Disabilities Act. Consonant with this disability perspective, Susan M. Wolf tells us, "feminists turn to the supposed biological fact of disability skeptically," so much so that "a feminist bioethics will have much to say about disability."[6] In the spirit of Wolf's remarks, I will suggest not only how feminist bioethics can incorporate a disability perspective, but also what it can contribute to the social model of disability.

In the first section, we will see why considerations of inclusiveness, intersectionality, and the danger of iterating women's historical oppression all urge that feminist biomedical ethics acquire a disability perspective. Next, I address key features of the standpoint of women with disabilities, including the oppression occasioned by the medicalization of disability, the consequent importance of disentangling disability from illness, and the concomitant identification of sources of suffering. The final section connects to the preceding discussions to show how feminist views can fortify a disability perspective in dealing with some so-called genetic dilemmas.

A Disability Perspective in Feminist Biomedical Ethics

Inclusiveness

There are several reasons for maintaining that feminist theorizing, and especially feminist biomedical ethics, should cultivate a disability perspective. A first and very substantial reason is that feminism is rightly admired for its commitment to embrace women who are especially disadvantaged. So, for example, Jaggar endorses "the sound moral intuition that in addressing moral and political problems, we must give the interests of more disadvantaged women special consideration."[7] And Wolf urges that "a feminist bioethics will pay special attention to subjects that emerge from the experiences of women who bear added burdens in dealing with . . . health care." She notes that "built into feminist method is the insistence that analysis start with attention to the individual, particularity and context." Thus "a feminist bioethics would not group all females under the rubric woman and proceed with a monolithic analysis."[8]

What difficulties lie in wait for those who attempt to introduce a disability perspective into feminism? Jaggar cautions us about the uneasiness occasioned by proposals to instill the concerns of women who are in the minority into majority feminist standpoints. Views that affect "different groups of women differently . . . are highly charged emotionally as well as morally" and "may involve incompatibilities among important feminist values or principles."[9] Jaggar also warns against permitting the fear of being "unfeminist" to derail consideration of women who are affected differently than most by the institutionalized mechanisms of domination.[10] This objection comes from those who urge that feminism must focus on securing gender equity. They question whether feminists have either the expertise or the mandate to address dimensions of inequality that are not gender specific.

The attitude Jaggar warns against results in an approach to disability much narrower than the one Wolf takes. On it, feminist biomedical ethics has nothing much to say about disability except to consider whether "the supposed biological fact of disability" promotes oppression that is disproportionately borne by women. On the understanding of feminism that informs this narrow view, for example, sterilization policy meant to reduce the births of people with disabilities is of concern only in respect to whether more women than men are sterilized. Feminist biomedical ethics' contribution thus is limited to the insistence that women should not be held primarily responsible for the transmission of disability. Feminism's prerogative and obligation are thought to lie in critiquing only the gender-differentiated implementation of such sterilization policy, not the gender-neutral repudiation of disability that underwrites it.

This assumes that we can dissect how women with disabilities experience oppression into gender-specific and disability-specific elements. Suffering exacerbated by multiple sources of bias cannot be so readily compartmentalized. So a second reason that feminism should cultivate a disability perspective is to enable theory to expand and comprehend better the standpoints of the many women whose experience lies mainly where various categories of oppression intersect.

Wolf reminds us that being identified with any minority group influences women's situations:

> Until recently, the key concerns in bioethics have been mainly dyadic. . . . This has obscured the importance of groups to which actors may belong. . . . Indeed, as data emerge showing correlations between health care and gender, as well as race and insurance status, it becomes all the more important to consider an individual's membership or perceived membership in a group.[11]

What of women simultaneously identified with more than one minority group? Feminist legal theorist Kimberlé Crenshaw maintains that an effective discourse of liberation must be both complex and compounded, capable of exposing and transfiguring several categories of oppression concurrently.[12] In commenting on the marginalization of women of color, she observes:

> they can receive protection only to the extent that their experiences are recognizably similar to those whose experiences tend to be reflected in antidiscrimination doctrine. . . . In order to include Black women, [feminism] must distance [itself] from earlier approaches in which experiences are relevant only when they are related to certain clearly identifiable causes (for example, the oppression of . . . women when based on gender) . . . the failure to embrace the complexities of compoundedness is not simply a matter of political will, but is also due to the influence of a way of

thinking about discrimination which structures politics so that struggles are categorized as singular issues.[13]

Women's color has affected how society permits them to participate in sexual activities; women of color have been assigned roles inferior to white women's. Crenshaw says: "Black women were assumed not to be chaste." She charges that feminism's "singular focus on rape as a manifestation of male power over female sexuality" is "an oversimplified account and . . . ultimately inadequate" because it ignores the institutionalized difference between white and black women. Thus, Crenshaw argues, an analysis of patriarchy rooted in white experience cannot address those aspects of intersectionality that pertain to roles in which black women have been devalued.

There are important similarities between what women with disabilities report regarding the pertinence of central notions of feminist theory to their experience and the concerns women of color like Crenshaw have raised. Both live in a general culture that denies to members of their groups roles sought by women and characters admired by women. The social construction of disability commonly prevents women with disabilities from assuming traditional female sexual roles. They are the group most likely to remain unmarried. Among people who have married and are not widowed, 12 percent of male nondisabled, 15 percent of female nondisabled, 11 percent of male disabled, but 25 percent of female disabled are divorced or separated. This data shows the sociocultural participation rate of women with disabilities descending below the combined straight-line projections of the participation rates of nondisabled women and men with disabilities, thereby suggesting that combined, the two stigmas have a more than additive negative effect. William Hanna and Elizabeth Rogovsky conclude: "Women in general, in contrast with men, are typically seen as having nurturing roles in our society. Our interviews suggest that women with physical disabilities are viewed differently . . . they are often seen as incapable of nurturing, indeed, as dependent people who must be nurtured."[14]

A fact sheet prepared for the Fourth United Nations World Conference on Women states:

> Women gave testimony as to how their disabilities had ended their marriages, isolated them from their families and communities, and destroyed their futures. Girls recounted how they were no longer regarded as future wives or mothers, but were instead hidden away from society. . . . Unlike other women, they have little chance to enter a marriage . . . which can offer a form of economic security. . . . Around the world, disabled women

are subjected to involuntary sterilization [and] pressured to, or required to, seek abortions.[15]

In this environment of exclusion, the differences of women with disabilities dominate how they are perceived and what experiences are permitted to them. "When I come into a room full of feminists, all they see is a wheelchair," reports disability activist Judy Heumann.[16] And Deborah Kent, blind since birth, reports:

> When I joined a women's consciousness-raising group a few years ago . . . I listened in amazement and awe as the others delivered outraged accounts of their exploitation at the hands of bosses, boyfriends, and passersby . . . it was impossible for me to confess my own reaction to their tales of horror, which was a very real sense of envy. . . . Society had provided a place for them as women, however restricting that place might be. . . . For myself and for other disabled women, sex discrimination is a secondary issue.[17]

In a study of how physically disabled young women in Sweden who have been the lifelong recipients of personal, parental, and medical care nevertheless seek their identity in traditional female caregiving roles, Karin Barron notes that "social life is organized around sex/gender and we generally behave consistently with the sex we were 'given' at birth."[18] She adds, "Instead of abandoning the mothering role, which has been described as a means of becoming 'equal' for women, disabled women may thus strive for this role . . . [but] this is not the traditional role of disabled women [who] have traditionally been denied the role of homemakers and mothers."[19] "Disability is a very powerful identity and one that has the power to transcend other identities. . . . For example, it has the power to de-sex people, so that people are viewed as disabled, not as men or women," observes disability scholar Tom Shakespeare.[20]

As Susan Sherwin remarks: "The contextual analysis sought by feminist ethics involves examination of the phenomenology and politics that arise from being assigned a position among the disabled in a society that demands perfection of its members."[21] We need to understand better how individuals can be banished to the margins of social participation and denied the roles customary for their gender by being identified as disabled. This discussion provides a third reason for urging a disability perspective on feminism.

An Iteration of Women's History

Today, disability is virtually identified with being physically, sensorily, or cognitively impaired, so much so that people in these very different conditions and with very different degrees of impairment are referred to, collectively, as "the disabled." Originally (and in legal discourse still today), having a disability literally meant having fewer, more limited, or more truncated rights than other classes of people. Thus to designate the class of individuals with physical, sensory, or cognitive impairments as "the disabled" is to regard them as ineligible for equal protection under the law. Their impairments defeat their appeals to rights, so to speak. It is because law or tradition deems them incompetent, or else their impairments bar them from meaningful exercise of the usual rights, that they have not had the usual protections.

In the nineteenth century, to be a woman was to have disabilities. For example, the Women's Disabilities Bill, debated in the British Parliament in the 1870s, addressed women's legal status rather than their physical or cognitive condition, although their characterization as a "weak," and thereby an inferior, class figured in rationalizing their unfavorable legal treatment. Women, regardless of their personal competence, were disabled from owning property, obtaining custody of their children, and voting. Such limitations were justified by characterizing women as belonging to a group of persons whom nature made too weak and stupid, too physically and morally frail, to execute business and head households successfully. Sherwin describes how medicalizing their differences contributed to women's disabilities:

> The mid-nineteenth century brought . . . a new medical interest . . . to establish menstruation as disability that demands rest and withdrawal from ordinary activities. . . . By the end of the century physicians were in the forefront of the campaign to . . . restrict women's participation . . . the prevailing . . . attitude was that menstruation created invalids out of women and made them particularly unfit. No thought was given . . . to adapting the demands of the universities or the workplace to these supposed special needs of women.[22]

Even though some women were demonstrably stronger and more competent than some men, their unequal status was defended as being in the state's interest. Society had to be protected from the typical woman's supposedly hapless attempts at independence and, concomitantly, had to protect the typically helpless woman from demands that supposedly would overwhelm her. This oppressive assessment of women's collective physical and emotional character is much less in evidence today, but it has neither been eliminated nor its deleterious influence fully erased.

We need not document at length how medicalizing elements of women's ordinary experiences as illnesses became a rationale for disempowering women by barring them from important social roles.[23] It suffices to note that a correspondingly oppressive history connects women's identification as a marginalized social group with disabled people's collective identity. There is a cultural dimension as well. For, despite having progressively liberated themselves from their former legal subordination, women, as are people with disabilities, are still disadvantaged by physical and social environments arranged to favor the physical and social preferences of men. Iris Marion Young cites this source of oppression when she attributes limitations in women's physical activity to the coercion of a patriarchal social structure that cannot abide to have women manifest full proficiency: "Women in sexist society are physically handicapped. Insofar as we learn to live out our existence in accordance with the definition that patriarchal culture assigns to us, we are physically inhibited, confined, positioned and objectified."[24]

Jaggar reports that "the Western tradition has consistently regarded women primarily as mothers and as sexual objects."[25] As women with disabilities commonly have not even been regarded as such, their experience is not addressed by analyses of women's oppression that foreground these roles. However, feminism can begin to address the intersection of gender and disability oppression by noting the striking resemblances in how the general culture has operated historically to debar both women and people with disabilities from flourishing. The medicalization of their physical, sensory, and cognitive differences has been central to the disempowerment of women. So, too, for people with physical, sensory, and cognitive impairments.

The Social Construction of Disability

Medical and Social Models of Disability

Disability researchers Brian Lamb and Susan Layzell comment:

> There is an unspoken taboo about relationships and disabled people. Disabled people's sexual and emotional needs are rarely included in any . . . representation in everyday life. . . . This reinforces the public's attitudes and expectations toward disabled people as seeing them as "sick and sexless" rather than as participating in full sexual and family relationships.[26]

There is ample precedent to establish the oppressiveness of characterizing people as ill, frail, and pathologically dependent when they are not so. The traditional conceptualization attributes the disadvantage associated with disability to individual defectiveness arising from illness, fragility, or infirmity. This medicalized model fixes on reducing the number of people with disabilities by preventive or curative medical technology and segregating those who remain uncured, putatively to focus on their special needs. But despite its scientific associations, the medical model has come more and more to resemble, in its assignment of responsibility, that antique belief that impairments are visited on individuals in response to their own or their ancestors' moral flaws. For, frequently, impaired individuals' deficits are attributed either to their own inadequate health or safety practices, or else to their bad genes, that is, to their own or their ancestors' defects.

Space is far too brief to present a thorough critique of the medical model of disability.[27] Nevertheless, some brief comments are in order.

First, the medical model of disability treats the built and arranged environment as an invariable to which humans have no choice but to adjust. However, it clearly is human to manipulate our environment. We fashion the built environment, which can be hostile or welcoming in respect to specific impairments. We, or at least the preeminent ones among us, also influence the dominant cooperative scheme that structures communication, allocation, modes of connectedness, and the other transactional processes of our social environment.

Second, some feminist theory recommends replacing the medical model of disability with a social model. As Wolf tells us:

A hallmark of feminist work has been its differentiation of the biological fact of sex from the cultural construction of gender. . . . Thus feminists turn to the supposed biological fact of disability skeptically. Martha Minow, for example, shows how disability and its effects are a function of social choice.[28]

Third, the 1990 Americans with Disabilities Act (ADA) is based on the belief that disability is socially constructed. The analysis of disability that informs this legislation merges a line of thought drawn from G. W. F. Hegel, Karl Marx, and Michel Foucault with the classical liberalism of the American civil rights movement. Problematizing disability this way began in the 1970s as a result of the influences of radical philosophy on the disability movement in Great Britain. Subsequently, U.S. disability activists adopted the model because it illuminated and gave a direction to social reform. The model explains the isolation of people with disabilities not as the unavoidable outcome of impairment, but rather as a product of stigmatizing social values and debilitating social arrangements.

In creating the Americans with Disabilities Act, the U.S. Congress found that historically, society has tended to isolate and segregate individuals with disabilities. Individuals with disabilities are a discrete and insular minority who have been . . . subjected to a history of purposeful unequal treatment, and relegated to a position of political powerlessness in our society . . . resulting from . . . assumptions not truly indicative of the . . . ability of such individuals to participate in, and contribute to, society.[29]

The thrust of this view is that disability is not a "natural kind," nor is the disadvantage attendant on it an immutable fact of nature. Once it is recognized that no biological mandate or evolutionary endorsement warrants the dominant group's fashions of functioning as optimally effective or efficient, we find that the main ingredient of being (perceived as) normal results from a social environment accustomed to people like oneself.

Central to having a disability perspective is recognition of how hostile environments, not personal deficits, disable people whose physical, sensory, or cognitive states are different from those of the dominant class. Disability studies scholar Liz Crow writes: "the social model of disability . . . gave me an understanding of my life . . . what I had always known, deep down, was confirmed. . . . It wasn't my body that was responsible for all my difficulties, it was external factors, the barriers constructed by the society in which I live."[30] From the standpoint of persons mobilizing in wheelchairs, for instance, disability is experienced not directly from their inability to walk, but indirectly from the feeling provoked in them by their exclusion from bathrooms, from theaters, from transportation, from places of work, and from access to medical services. If the majority of people wheeled rather than walked, graceful spiral ramps instead of jarringly angular staircases would connect lower to upper floors of buildings. Were vision-impaired individuals dominant, information would not be conveyed in a format accessible only to the sighted; tactile and aural modes of recording information would be used as frequently as printed texts. The deficit customarily associated with disability is neither more nor less than an alterable cultural artifact.

Illness, Disease, Disability

To employ the social model of disability, it is crucial that disability be sufficiently disconnected from illness and pain so as to ensure that no judgment about the lives of people with disabilities is distorted by uncritical assumptions about their suffering. We may begin by noticing how ordinary language marks the difference between illness and disability. We speak of suffering an illness but of having or living with a disability. We cannot be both well and ill at the same time, but we can be both perfectly well and yet disabled, as are the competitors in the Para-Olympics. Ron Amundson insists forcefully on preserving this distinction in his 1992 essay "Disability, Handicap, and the Environment."[31] He offers three arguments.

First, persons with paradigmatic disabilities—paraplegia, blindness, deafness, and others—neither require nor are improved by medical treatment, differentiating them from people suffering from illnesses. When we are ill, we take medication to relieve pain, discomfort, weakness, or loss of appetite, but there are no medications for disability. Its remedy lies in social reform, not in biochemical intervention.

Second, the distinction can be sharpened by noticing what shifts when illness (or accident) results in disability. Although illness is traceable to specific physiological sites, its effects are diffused and so are manifested not just in pathology at the site, but in such conditions as pain, discomfort, weakness, lassitude, absence of appetite, and disorientation. Illness prohibits performances that have no direct association with the impaired site(s), and this global debilitation invites the individual who is ill to assume the "sick role." Disabilities, conversely, rarely involve any such diffused manifestations.

Third, disability usually involves delimited and related areas of functioning, while illness is more globally debilitating. Thus, we describe someone who cannot see as being "visually disabled," meaning she is unable to engage in performances that necessitate vision (although she may successfully execute such functions as reading, that are typically performed by seeing, by adopting alternative performance modes). Or we call the group of people who cannot manipulate text (because they cannot see it, cannot interpret it, or cannot hold the pages or papers on which it is inscribed) "print disabled." In contrast, an individual who is ill usually is more globally and diffusely incapacitated.

As Amundson points out, the "sick" role is a kind of social stepping or stopping out that is inappropriate for someone with no illness but only a disability.

> The 'sick role' . . . relieves a person of normal responsibilities, but carries other obligations with it. The sick person is expected to . . . regard his or her condition as undesirable. These requirements resonate with the attitudes of society towards disabled people. . . . One interesting correlation is that able bodied people are often offended by disabled people who appear satisfied or happy with their condition. A mood of regret and sadness is socially expected.[32]

None of this is to deny that illness can be the preceding or chronic cause of disability (although traumatic accident is equally a cause of disability). As Susan Wendell points out, people often are disabled by chronic diseases such as diabetes that cause a variety of impairments, including loss of limbs and blindness.[33] Also, the hostility of the environments people with disabilities must endure sometimes causes them to become ill. For instance, an environmental defect such as the absence of wheelchair-accessible bathrooms increases bladder and kidney disease in paraplegics, who are barred from the physiological benefit of relieving themselves by buildings designed to accommodate the general public but not people using wheelchairs. The fact that it is informative to say diseases cause disabilities or increase one's likelihood of becoming disabled indicates that there is no analytic, definitive, or conceptual connection or identity between the two conditions.

Amundson is not the only writer who recognizes the disadvantage association with sickness imposes on those who are not ill, who need no recovery period, and who are harmed rather than benefited by the disruption of or the distancing from everyday responsibility. As we come to understand genetic structures accurately enough to identify the potentially anomalous modes and levels of functioning consequent on any species' member's genetic inheritance, Dan Brock warns, some members of the species will find themselves devalued by their own futures. Although performing splendidly at the time, they will be marginalized because they are at greater genetic risk than others of deteriorating function. So, for instance, those at increased risk of Alzheimer's might be rejected as mates by those who fear future service as a spousal caretaker, while employers desiring to keep medical insurance costs down might not hire individuals genetically disposed to develop various kinds of cancer. Merely being connected to prospective illness may well damage their well-being and self-esteem:

> Generally it is when we have noticed an adverse effect or change in our normal functional capacity that we contact health care professionals and begin the process which can result in our being labeled as sick or diseased . . . people who feel healthy and who as yet suffer no functional impairment will increasingly be labeled as unhealthy or diseased (by genetic testing). . . . For many people, this labeling will undermine their sense of themselves as healthy, well-functioning individuals and will have serious adverse effects both on their conceptions of themselves and the quality of their lives.[34]

Who Is Disabled?

Wendell fears that "some of the initial opposition in disability rights groups to including people with illnesses in the category of people with disabilities may have come from an understandable desire to avoid the additional stigma of illness."[35] Her point is well taken, and it is important to notice that the distinction I draw does not deny that illness can result in disability. Nor is there a justification for excluding people from being identified as disabled because they are ill. But being ill, even chronically ill, is neither required for, nor decisive evidence of, being disabled.

Who is designated as disabled has commonly been determined by the eligibility criteria for programs that offer charitable treatment, special public benefits, or exemptions from obligations to members of specific groups that different charitable policies are designed to serve. Programs vary in their definitions. However, because our purpose is to examine the intersection between gender and disability oppression, the definitions established by such entitlement programs are of less relevance, because narrower, than the one designed to identify disability discrimination in the 1990 ADA. In this legislation, disability means the substantial limitation of one or more major life activities due to a physical or mental impairment, or having a history of such, or being regarded as having such. To illustrate, neither retinal damage nor being in pain are, on this understanding, disabilities. They are relevant to disability only to the extent that they substantially limit such major life activities as walking or seeing.[36]

The ADA establishes a generally applicable civil right rather than a group-differentiated right. It assigns all individuals the right to be protected against disability discrimination. But only some individuals can claim group-differentiated disability rights to benefits, entitlements, insurance, and compensation. Unlike the definitions of disability created for entitlement programs, all of which reflect prior agreements regarding the ultimate size and scope of the clientele to be served, the ADA contains no notion of eligibility other than the demonstration that an individual has been discriminated against because she is believed to have an impairment.

Unlike protection from disability discrimination, group-differentiated disability rights are not corollaries of democratic principles. Disability entitlements, benefits, insurance, and compensation are instances of special treatment, not equal treatment, and are the products of programs created for the purpose of sustaining and managing groups of people considered too weak, vulnerable, incompetent, or damaged to fend for or support themselves.[37] Each program defines for itself what makes individuals eligible for benefits. For instance, workers' compensation programs define disability as dysfunction related to the performance of employer-assigned tasks, whereas disability is given a notably different meaning for the purposes of various publicly financed social welfare programs.

While these definitions focus on what individuals cannot do, the ADA addresses how they are competent. This is because disability discrimination is the misperception that being limited in performing major activities also limits competence. Individuals substantially limited in one or more major life activities are protected by the ADA against the hostility of built or arranged environments that exclude them from opportunities they are otherwise competent to pursue. The exclusion of individuals because they are mistakenly regarded as being so limited also is prohibited.

Congress heard extensive testimony from people with many kinds of disabilities, women and men, young and old, of every economic class and situation. The provisions of the ADA reflect the legislature's conviction that what is common to individuals who experience disability is neither incompetence nor dependence nor any special needs, but instead a history of exclusion that disregards and devalues the facts of their functionality and competences. When women are denied opportunity because the physical or mental characteristics of their sex are mistakenly believed to make them physically or mentally unfit for certain functions, the injustice done to them is analogous to disability discrimination.

Suffering

When exclusion on the basis of disability permeates one's experience, the characteristic disability perspective or standpoint takes shape. Political scientist Harlan Hahn's description articulates what many individuals with long-standing disabilities understand to be the experience of disability:

> One of the most unpleasant features of the lifestyles of . . . disabled individuals . . . [is] the pervasive sense of physical and social isolation produced not only by the restrictions of the built environment but also by the aversive reactions of the nondisabled that often consign them to the role of distant friends or even mascots rather than to a more intimate status as peers, competitors, or mates. Few nondisabled individuals would tolerate the curtailments of individual options that become part of the daily experience of people with disabilities.[38]

It is this deprivation of connectedness that ties disability to suffering. A key difference between the medical and the social model of disability in this regard lies in how each proposes to remedy the misalignment between individual and social practice. The social model does not deny that to have a disability is to be at heightened risk of suffering, but in attributing this liability, it gives first consideration to the hostile environment that rebuffs the disabled individual when she tries to perform under the conditions it imposes.

It is sometimes claimed that much of the suffering disabled persons endure is the result of a society that fails to care enough to support those who must depend on community resources more than others commonly do. However, we currently are not able to assess such claims about the cause of their suffering because we have no idea what percentage of people with disabilities would flourish with no special call upon community resources if discrimination against them were eliminated. The exclusion Hahn describes as pervasive in their lives has so reduced their opportunities for social participation as to render them artificially dependent when they need not be so.

We should also recognize that both public and private charitable support programs often insist that individuals with disabilities are more dependent than nondisabled people in order to segregate them from the general population. Paratransit systems, special education classrooms, and exemptions from paying fees, all of which are used to justify not making facilities fully accessible, keep people with disabilities separate from and unequal to the general public. Separate facilities usually cost more to maintain (if they are equal) than integrated ones. They also impede forming community and family bonds.

The social model makes clear that working and living conditions first must be revised to provide equal access before the productivity of individuals with and without disabilities, their real dependence or needs, and their propensity for suffering can be fairly assessed. Pending evidence obtained under less biased conditions than presently obtain, we should avoid assuming that compromised productivity, neediness, and unusual suffering attach to the disabled with any strength other than that resulting from the exclusionary practices feminists should recognize as reiterating the familiar forms of women's oppression. The virtue of such caution is illustrated by Elizabeth Spelman in her book *Fruits of Sorrow*: "Claims of . . . human suffering can do as much to reinforce claims of superiority and inferiority as they can to undermine them."[39]

Applications at the Intersect

A Feminist Perspective on the Social Model

Even under current conditions, individuals with the same kind and degree of impairment attain vastly differing levels of social success. To illustrate, deaf people can be successful musicians (Evelyn Glennie); blind people can be outstanding scientists (Geerat Vermeij);[40] and although mobility-impaired people commonly are thought to be unfit for strenuous work, some have thrived in almost every sort of employment (rock climber, war correspondent, tennis player, pilot) including the presidency of the United States (Franklin D. Roosevelt). The variables most likely to explain why identically impaired individuals achieve so differently are (1) individual talent or lack of it, (2) effort expended, and (3) opportunities offered in an impairment-friendly environment or eliminated by an impairment-hostile one.

While acknowledging that individual variations in effort and talent will yield different outcomes and different degrees of success, democratic political morality nevertheless urges reshaping the social environment so as to create more uniformly accessible opportunity. The social model of disability presses for such changes in social arrangements, that is, for transformations in what Allen Buchanan calls "the dominant cooperative scheme."[41] Buchanan points out that choosing the dominant cooperative scheme means choosing who will and will not be disabled because those whose preferred modes of functioning are facilitated by the arranged environment will not be dysfunctional. And whoever imposes an environment that suits themselves will disable those who function substantially differently.

However, Buchanan does not agree with the recommendation that we should reform society, not revise individuals, so as to equalize individuals' opportunities for whatever achievement their talents can effect. Instead, he argues, each individual has an interest in establishing the cooperate scheme in which he can be most productive, even if so doing excludes others who are not productive under this scheme. Furthermore, arrangements that accommodate individuals with physical, sensory, or cognitive impairments are not those that stimulate normal individuals, who, he imagines, flourish in a more challenging environment than do individuals with disabilities. He says, "participation by 'disabled' individuals can cause discoordination, and can reduce the benefits which the 'abled' might otherwise obtain from the form of interaction in question."[42] Consequently, although we all have an interest in inclusion, Buchanan thinks the majority also has an interest in creating an environment that fits its own modes of functioning most productively. Buchanan believes this environment cannot help proving exclusionary for people with disabilities.[43]

Buchanan says, "We must reject the disabilities rights advocates' slogan that we should 'change society, not individuals.' "[44] While "some compromises with efficiency are required in the name of equal opportunity for those with disabilities . . . the interests of employers (and workers who do not have disabilities) are also legitimate," so "a just effort to achieve a better fit between individuals' abilities and the demands of the dominant cooperative scheme may require changing individuals through genetic means or otherwise."[45] On this view, the dominant cooperative scheme appears to be invariable. Instead of reforming the scheme, we will have to retrofit individuals if social and economic participation are to be distributed more widely. It is comments like Buchanan's that prompt Adrienne Asch and Gail Geller to worry that "the Human Genome Initiative could turn out to make 'species-typical functioning' a guide to joining or remaining part of the human community."[46]

Here feminist theory can correct Buchanan's assumption about what makes a cooperative scheme productive. For it is oversimplified to think that what is in contention between those who seek inclusion for the disabled and those who would exclude them is the preservation of efficiency, or the most productive system. There is no reason to agree with Buchanan that the practices of our current competitive and exclusionary scheme are optimally productive.

Feminist social science is replete with studies that demonstrate the waste and harm of our current "not very cooperative" scheme. Consequently, feminists Hilde Lindemann Nelson and James Lindemann Nelson recommend our sharing "the feminist suspicion regarding prevailing social institutions. . . . Feminist theorists . . . will have little reason to take those structures quite as seriously as theorists who are comfortably at home inside them."[47] We are reminded again of why feminists should forbear from imagining what kinds of people need more or are more dependent on community resources than "normal" people. The need for special provisions is itself often an artifact of prevailing social institutions that elevate the convenience (not the productivity) of the kinds of individuals who are dominant into a rigidly exclusionary standard.

Buchanan's application of the social model of disability invites feminist criticism of his entrepreneurial point of view. He believes that to satisfy the interests of those who think social arrangements must be maximally challenging to be maximally productive, it is appropriate to disadvantage whoever cannot rise to such strenuous challenges: "The problem of justice is this: satisfying the interest in inclusion imposes costs on those who could benefit from a more demanding scheme; satisfying the maximizing interest imposes costs on those who would be excluded from effective participation in it."[48] But to portray the interests of the nondisabled and the disabled as pitted against each other in this way does not fairly represent the situation. Approaches such as those identified by Nelson and Nelson help here: "Like other feminists we maintain a skepticism . . . about the unrealistically atomistic picture of human relationships that is presumed in the social contract tradition."[49]

First, Buchanan ignores those more optimal alternatives to the current dominant scheme on which the nondisabled are no less productive and the disabled become more productive. Nor does he demonstrate the unfeasibility of such alternatives. Second, Buchanan wrongly suggests that few individuals with a disability are as efficient and as ready for challenges as other people are, a belief that shows he remains in the thrall of the medical model of disability. Third, this portrayal of competing interests overlooks the complexities of our social interdependence.

Buchanan speaks as if each individual's efficiency taken separately is all that matters in increasing productivity. But complex social factors, including how workers are stimulated by one another's well-being, affect productivity. If we promote connectedness as vigorously as we do individual competition, feminism suggests, schemes that respond to a diversity of modes of functioning can amplify productivity. Moreover, feminist social philosophy urges us to pursue ways of serving one person's interests that further other people's interests as well.

Feminist theory helps us to understand why those who flourish under the dominant scheme have, like those who do not, an interest in fashioning more inclusive practices. For as the U.S. Congress observes in the "Findings" section prefacing the ADA, "the continuing existence of unfair and unnecessary discrimination and prejudice denies people with disabilities the opportunity to compete on an equal basis . . . and costs the United States billions of dollars in unnecessary expenses resulting from dependency and nonproductivity."[50] Inclusion lowers productivity, Buchanan worries, not noticing that exclusion, too, lowers productivity. After all, the more unproductive people there are, the greater the drain on the communal resources to which all who are productive contribute.

A Case Study: Desiring Deaf Children

Feminist theory has yet another contribution to make to strengthening the social model of disability. As Nelson and Nelson put it, "feminism expands moral vision, offering a way of seeing otherwise obscured injustices."[51] To illustrate how such injustices may be obscured in considering what is appropriate health care for women with disabilities, I now turn to a predicament in genetic counseling that appears to pit the interests of parents with disabilities against their future children.

In the *Hastings Center Report,* Dena S. Davis explains that parents with disabilities who desire a child with their same disability precipitate, in her words, a genetic dilemma for counselors:

> Deeply committed to the principle of giving clients value-free information . . . , most counselors nonetheless make certain assumptions about health and disability—for example, that it is preferable to be a hearing person rather than a deaf person. Thus, counselors typically talk of the "risk" of having a child with a particular genetic condition . . . parents with certain disabilities who seek help in trying to assure that they will have a child who shares their disability . . . are understandably troubling to genetic counselors.[52]

Davis adopts the social model of disability in pointing out that deaf individuals function competently in the Deaf community but are disabled by the practices of the nonsigners who dominate the hearing community. (The uppercase "D" indicates people who are culturally Deaf as well as those who are physically deaf.) To Davis, this means that to be deaf is to be denied a choice of communities to join. The "narrow choice of vocation" deaf people have is, she thinks, "likely to continue to lead to lower standards of living. (Certainly one reason why the [Martha's] Vineyard deaf were as prosperous as their neighbors was that farming and fishing were just about the only occupations available)."[53] Davis argues that a child's right to an open future is violated if the choices the child can make as an adult are narrowed by decisions made by others prior to birth. A liberal state "must tolerate even those communities most unsympathetic to individual choice." but must preserve "the right of individuals to choose which communities they wish to join or leave," she says.[54]

Davis assumes, crucially, that the desire to impose their own values on their children motivates Deaf parents to seek deaf children: "Deaf parents wishing to ensure Deaf children are an example of families . . . shaping . . . a radically narrow range of choices available to the child when she grows up. . . . Liberalism requires us to intervene to support the child's future ability to make her own choices about which of the many diverse visions of life she wishes to embrace."[55] If deafness is considered a disability, one that substantially narrows a child's career, marriage, and cultural options in the future, Davis concludes, then "deliberately creating a deaf child counts as a moral harm. . . . The very value of autonomy that grounds the ethics of genetic counseling should preclude assisting parents in a project that so dramatically narrows the autonomy of the child to be."[56]

Feminist theory encourages us to wonder whether Davis's account of the benefits of autonomy abstracts unrealistically from the experiences of Deaf parents and ignores the interdependence of parents and children. Are these Deaf mothers, as Davis thinks, more concerned with the homogeneity of their family than with the child's well-being? I believe that Davis's conceptualization of the motivation of Deaf mothers is compromised by our culture's reluctance to imagine women with disabilities as being fully functional in parenting roles.[57] To be fair to Deaf mothers, we should assume that they, like hearing mothers, are motivated by concern for their prospective children.

Davis portrays genetic counselors as having no principled objections to assisting nondisabled parents to select nondisabled over disabled progeny if they feel unable to parent a child with a disability. The decisive consideration is expressed by a counselor she quotes: "I am not going to be taking that baby home—they will."[58] Surely, then, counselors should give the same benefit of doubt to Deaf mothers that they do to hearing ones. If Deaf mothers think they can better nurture deaf than hearing children, shouldn't they have the same assistance from counselors as hearing mothers who think they can better nurture hearing than deaf children?

A child's entitlement to the best possible parenting counts heavily with genetic counselors when they assist hearing mothers. Are not the children of Deaf mothers equally entitled to be well matched to their family's parenting skills? Hearing adults who judge themselves incompetent to parent deaf children do so because they are not expert in how communication develops and how educational and career paths unfold for such children. So it is understandable that Deaf mothers, who know the roads deaf people travel to secure a good education and career better than they do the paths of the hearing world, might well be more confident in guiding a child along the paths they know, the paths of Deafness. Parenthetically, Davis reports that Deaf children of Deaf parents are "the academic creme de la creme," which suggests the effectiveness of good Deaf parenting of Deaf children.[59]

Granted, deaf people are a minority in the hearing world, and society currently gives them fewer career options than hearing people. Feminists suspicious of existing institutional arrangements might point out that how substantial this difference seems is a matter of perspective. Being deaf does not foreclose legal, scientific, academic, acting, and music careers, among many others. Because one has only one life to plan, the ultimate benefit to the child's autonomy of having fewer life plan options from which to choose may be negligible, more like having a varied menu with ten rather than twelve main courses than like being a vegetarian in a steak house.

We should not identify having a greater number of options with the meaningful exercise of autonomy, the value Davis wants genetic counselors to preserve for children's futures. Many feminists have pointed out that autonomy should not be conflated with atomistic individualism. Moreover, as Diana Tietjens Meyers shows very persuasively, to act autonomously is not necessarily to have complete control of one's own destiny.[60] At least as important to well-being as having many options is whether parents can prepare their children to overcome disadvantageous circumstances. A feminist perspective reminds us that the advantage of having a few more options may be exaggerated; just as important are the many interconnected family strengths that encourage a child to flower.

Indeed, having too many opportunities can compromise a child's future. Uncertainty about which road to take and regret about roads open but not taken are familiar life spoilers. Children often do better if they focus themselves on a few satisfying options instead of being torn with indecision by many glittering ones. To "open" a child's future, then, is not just to offer an array of life plan options. At least as important is to teach the child, by example as well as by principle, to make good choices in a world where not everything is possible. To do so is surely to facilitate rather than to compromise the child's future well-being.

Nor need a reduction in options reduce a child's prospect of thriving. If it did, all parents proposing to have children in disadvantageous circumstances should in principle be discouraged lest they diminish their child's future flourishing. This is far too strong a thesis, for it could be extended to deter parents from having female children in societies inimical to their sex or to coerce single mothers to give their children up for adoption by two-parent families.

This raises the question of when it is morally wrong to select a child's inherited characteristics. We cannot facilely condemn all such decisions, for in choosing our children's biological fathers, consideration of heritable traits—intelligence, strength, and so on—is not easily suppressed. Davis opposes selections that impose parental values in a way that curtails the child's own choices. Selecting the sex of one's child may be as dangerous to the child as selecting a child who does not hear: "Parents whose preferences are compelling enough . . . to take active steps to control the outcome . . . are likely to make it difficult for the actual child to resist their expectations."[61]

Mothers who practice sex selection in cultures hostile to the decent treatment and advancement of women often say their choice is born from despair at the suffering daughters would have to endure. But many feminists for whom confronting the oppression of women is paramount insist that we defeat ourselves if we permit flawed social arrangements to cause us to depreciate either ourselves or our prospective offspring made in our own image. Consequently, we should not reject (prospective) people because they are female, even if doing so is spurred by a realistic assessment of the bias they will face. This feminist basis for objecting to sex selection is very different from Davis's reason.

Inflected with a disability perspective, this feminist basis also generates opposition to counseling practice that, in its disrespect for deaf people, resembles the hostility of male-venerating cultures that drive mothers to fear having female children. Given their group history, it should be no surprise that, for Deaf parents, the hearing counselors' assumption that "it is preferable to be a hearing person than a deaf person" reflects the well-known social bias against recognizing the value of deaf people's lives. Feminist theory advises contextualizing this assumption to ascertain whether or not it theorizes an oppressive practice, one that seizes upon their difference to justify the inferior treatment of deaf people.

Adopting a disability perspective enables us to understand Deaf mothers as preferring to have deaf children not because they want to deny their offspring's autonomy, but because they need to secure their own parenting against an environment they perceive to be hostile. For instance, children whose parents have disabilities are often belittled by their playmates. As Paul Preston reports in *Mother Father Deaf: Living Between Sound and Silence,* hearing children of deaf parents envy deaf siblings who can remain unaware that their families are the subject of loud ridicule.[62] That hearing children of deaf parents must share the social penalties of being deaf afflicts the parents and embitters their hearing children.

Deaf mothers' perspectives emerge from their knowledge of the social barriers women with disabilities face in trying to parent. It has not been unusual for deaf parents to lose their hearing children to a social service system that denies them access through qualified sign-language interpreters to the adjudicative "hearing" processes. This sad history continues to this day. To fully appreciate the hazards the deaf mothers perceive, one must understand that depriving the children of deaf parents of a signing environment by placing them in foster care during their initial language learning compromises their communication with their parents. These children then find themselves distanced from their parents by circumstances similar to those distancing deaf children whose hearing parents either don't sign or sign badly. Deaf mothers seeking deaf children are consistently acting on a principle equally applicable to hearing mothers: act as to maximize communication with your child.

Inflecting feminist theory with a disability perspective permits us to

appreciate mothers with disabilities who seek to implement this principle and suggests how to weigh the propriety of mothers without disabilities seeking children without disabilities. Disability activists who
adopt an emotive version of the social model decry genetic counseling
that results in children with disabilities not being born. They then condemn this use of genetic counseling, saying it expresses negative feelings about all people with disabilities, a conclusion that does not follow, as Buchanan correctly argues.[63] The position they take, Oliver
explains, ascribes the oppression of people with disabilities to sheer social hatred of them, rather than, more plausibly, to specifically oppressive forms of practice that debar them from assuming productive social
roles.[64]

This analysis has shown the genetic dilemma Davis describes to be
precipitated not by the prospect of future children with disabilities but
by the practice of giving the judgment of prospective parents less
weight if they have a disability. Usually, genetic counselors feel bound
to give their clients the information they need to make their own
choices, Davis says.[65] Counseling practice begins to go wrong when
nondisabled mothers' judgments about whether they can "risk" parenting a child with a disability get more respect than disabled mothers'
judgments about whether they can risk parenting a child without a disability.

In short, prospective mothers who are deaf are disempowered by
counselors who do not appreciate the standpoint of deaf people. Counseling practice here fails to respect these prospective mothers. The
practice is biased by presuming that these mothers are not competent
to protect their own children's welfare. In other words, their impairment is imagined to so weaken their judgment that they are not fully
qualified to execute their discretionary parental role. What is wrong
here is not abhorrence of people with disabilities, but a phenomenon
long familiar to women, namely, that same denial of competence and
the consequent disempowerment of women that, historically, has been
women's disability.

Notes

1. Quoted in Alex Witchel, "Only in the Kitchen Are There No Letdowns,"
New York Times, July 23, 1997, B9.
2. Alison Jaggar, "Introduction," *Living with Contradictions: Controversies in
Feminist Social Ethics* (Boulder, CO: Westview Press, 1994), 11.
3. Virginia Kallianes and Phyllis Rubenfeld, "Disabled Women and Reproductive Rights," *Disability & Society* 12, no. 2 (Spring 1997): 203.

4. Anita Silvers, "Women and Disability," in *Blackwell's Companion to Feminist Philosophy*, ed. Alison Jaggar and Iris Marion Young (Oxford: Blackwell's, 1998).

5. Michael Oliver, "Defining Impairment and Disability: Issues at Stake," in *Exploring the Divide: Illness and Disability*, ed. Colin Barnes and Geof Mercer (Leeds: Disability Press, 1996), 39.

6. Susan M. Wolf, "Introduction," in *Feminism and Bioethics: Beyond Reproduction*, ed. Susan M. Wolf (Oxford: Oxford University Press, 1996), 23

7. Jaggar, *Contradictions*, 10.

8. Wolf, *Feminism*, 23.

9. Jaggar, *Contradictions*, 8.

10. Ibid.

11. Wolf, *Feminism*, 17.

12. Kimberlé Crenshaw, "Demarginalizing the Intersection of Race and Sex: A Black Feminist Critique of Antidiscrimination Doctrine, Feminist Theory, and Antiracist Politics," in *Contradictions*, ed. Jaggar, 39–52.

13. Ibid., 48.

14. Bureau of the Census, *Survey of Income and Program Participants*, SIPP84-R3 (Washington, DC: Department of Commerce, 1984).

15. Rehabilitation International and World Institute on Disability, "Fact Sheet" prepared for the United Nations Fourth World Conference on Women. (New York and Oakland, CA: World Institute on Disability and Rehabilitation International, July 1995; updated 1997).

16. Quoted in Rosemary Garland Thomson, *Extraordinary Bodies: Figuring Physical Disability in American Culture and Literature* (New York: Columbia University Press, 1997), 26.

17. Deborah Kent, "In Search of Liberation," in *Disabled USA* 1, no. 3 (1977).

18. Karin Barron, "The Bumpy Road to Womanhood" in *Disability & Society* 12 no. 2 (Spring 1997): 224.

19. Ibid., 234.

20. Tom Shakespeare, "Disability, Identity, and Difference," in *Exploring the Divide*, ed. Barnes and Mercer, 109.

21. Susan Sherwin, *No Longer Patient* (Philadelphia: Temple University Press, 1992), 91.

22. Ibid., 182.

23. For an excellent summary, see Sherwin's *No Longer Patient*.

24. Iris Marion Young, *Throwing Like a Girl and Other Essays in Feminist Philosophy and Social Theory* (Bloomington: Indiana University Press, 1990), 153.

25. Jaggar, *Contradictions*, 7.

26. Brian Lamb and Susan Layzell, *Disabled in Britain: A World Apart* (London: SCOPE, 1994), 21

27. Anita Silvers, "(In)Equality, (Ab)normality, and the 'Americans With Disabilities' Act," *Journal of Medicine and Philosophy* 21 (1996): 209–224; "Damaged Goods: Does Disability Disqualify People From Just Health Care?" *Mount Sinai Journal of Medicine* 62, no. 2 (1995): 102–111; and "Disability Rights", *The Encyclopedia of Applied Ethics*, ed. Ruth Chadwick (San Diego: Academic Press, 1997), vol. I, 781–796.

28. Wolf, *Feminism*, 23

29. Public Law 101–336, *Americans with Disabilities Act of 1990*, sect. 2 (7).

30. Liz Crow, "Including All of Our Lives: Renewing the Social Model of Disability," in *Exploring the Divide*, ed. Barnes and Mercer, 55–73.

31. Ron Amundson, "Disability, Handicap, and the Environment," *Journal of Social Philosophy* 23, no. 1 (1992).

32. Ibid., 114, 118.

33. Susan Wendell, *The Rejected Body* (London: Routledge, 1996), 20.

34. Dan Brock, *Life and Death* (Cambridge: Cambridge University Press, 1993), 29.

35. Wendell, *Rejected Body*, 21.

36. ADA, Sect. 3 (2).

37. Silvers, "Disability Rights."

38. Harlan Hahn, "Civil Rights for Disabled Americans," in *Disabling Images* ed. Alan Gartner and Tom Joe (New York: Praeger, 1987), 198.

39. Elizabeth Spelman, *Fruits of Sorrow: Framing Our Attention to Suffering* (Boston: Beacon Press, 1997), 9.

40. Geerat Vermcij, *Privileged Hands: A Scientific Life* (New York: W. H. Freeman, 1997).

41. Allen Buchanan, "Choosing Who Will Be Disabled: Genetic Intervention and the Morality of Inclusion," *Social Philosophy and Policy* 5, no. 13 (Summer 1996): 40.

42. Ibid., 41.

43. Ibid., 44.

44. Ibid., 45.

45. Ibid., 42.

46. Adrienne Asch and Gail Geller, "Feminism, Bioethics, and Genetics," in *Feminism and Bioethics*, ed. Susan M. Wolf, 318–350.

47. Hilde Lindemann Nelson and James Lindemann Nelson, "Justice in the Allocation of Health Care Resources: A Feminist Account," in *Feminism & Bioethics*, ed. Susan M. Wolf, 354.

48. Buchanan, "Choosing Who," 42.

49. Nelson and Nelson, "Justice," 354.

50. ADA, sect. 2 (a9).

51. Nelson and Nelson, "Justice," 354.

52. Dena S. Davis, "Genetic Dilemmas and the Child's Right to an Open Future," *Hastings Center Report* 27, no. 2 (March–April 1997): 7.

53. Ibid., 13. The reference to Martha's Vineyard refers to the fact that, due to the prevalence of hereditary deafness in that community, for more than a century hearing as well as deaf residents conversed by signing.

54. Ibid., 9.

55. Ibid., 11.

56. Ibid., 14.

57. Rosemarie Garland Thomson reveals much about this bias in *Extraordinary Bodies*, her recent study of how American literature portrays women with disabilities.

58. Davis, "Genetic Dilemmas," 8.

59. Ibid., 13.

60. Diana Tietjens Meyers, "Personal Autonomy and the Paradox of Feminine Socialism," *Journal of Philosophy* 84, no. 11 (November 1987): 622.

61. Davis, "Genetic Dilemmas," 14.

62. Paul Preston, *Mother Father Deaf: Living Between Sound and Silence* (Cambridge, MA: Harvard University Press, 1994).

63. Buchanan, "Choosing Who," 28–33.

64. Oliver, "Defining Impairment," 50.

65. Davis, "Genetic Dilemmas," 7.

11

Menopause: Is This a Disease and Should We Treat It?

Wendy A. Rogers

enopause is frequently referred to as "estrogen deficiency disease" in the medical literature.[1] This assertion is coupled with the further allegations that postmenopausal osteoporotic fractures exact a huge cost on the community (Wark 1996) and that the medical profession should overcome this disease with hormone replacement therapy (HRT). The story linking these two points is as follows: menopause and the postmenopausal state are diseases, as menopause is associated with unpleasant symptoms and is followed by significant endocrine abnormalities. The effects of these diseases include increased risk of cardiovascular disease and of osteoporosis. The latter results in fractured hips, which, in Australia alone, cost somewhere between $240 and $389 million in Australian dollars (Aus$).[2] HRT would fulfill the dual functions of returning women to a normal estrogenic state and saving the community the cost of treating osteoporosis-related fractures.

My aim in this chapter is to critically examine claims in this medical story: first, that menopause is a disease; and, second, that women should take HRT to prevent the costly consequences of menopausally related osteoporosis. These claims warrant our attention, not least because of the normative power of medicine in shaping our cultural responses to menopause. Just as culture influences our understanding of medicine, medicine in turn influences the ways that we construct our identities within that culture: "Our perceptions of menopause are tied to the broader culture's underlying assumptions about womanhood, aging and medicine in general" (Barbre 1993, 24). Significantly, medicine is associated with judgments that are simultaneously informed by selected norms of human functioning and characterized as objective matters of fact (Margolis 1981).[3] However, both cultural assumptions and the basis for the medical claims are questionable.[4] The first section of this chapter briefly investigates the nature of disease. A detailed discussion of menopause follows, in which two competing accounts of menopause as disease are compared. The final section explores the arguments for widespread prescription of HRT to menopausal women.

The Nature of Disease

What do we mean by "disease," and how does medicine understand concepts such as "normal" and "abnormal" in relation to disease? The concept of disease is used to account for pathological or psychological disorders, offering generalizations about patterns of phenomena that are in conflict with our ideals of health. Such a concept is useful as an organizing concept, allowing us to explain, control, and predict various symptoms. A set of symptoms comprising fever, headache, and neck stiffness, for example, once classified as bacterial meningitis, becomes a disease with a known cause, course, treatment, and prognosis. As an organizing concept, "disease" functions as a straightforward descriptive term; however, this descriptive term encompasses significant normative components.

The first normative component of disease is statistical: disease is a variation from a statistical physiological norm (King 1981, 110; Boorse 1981, 546). Thus hypertension is a variation along the spectrum of normal blood pressure; diabetes an elevation of the glucose normally found in the blood; osteoporosis a decrease in the calcium levels found in normal bones. Physiological norms are derived from measuring a large number of parameters in a population and converting these into tabular forms with cutoff levels for "normal." Usually, the limits of normal are defined as lying within two standard deviations of the mean. Anything outside this is by definition abnormal, irrespective of whether or not the person with the abnormal values feels ill.

Prima facie, this is a purely descriptive exercise, with "normal" being equivalent to "average." However, there is a significant lack of factual rigidity in the measurement of normality. No bodily parameter can be interpreted as an isolated, objective fact but is always relative to individual conditions; it is a hemoglobin level in a certain person, of a certain gender, of a certain age, with or without other relevant conditions. An apparently abnormal level may be recognized as normal for a given situation: a hemoglobin count normal in pregnancy, for instance, would be considered anemia in a nonpregnant woman; similarly, the normal hemoglobin level of a person with renal disease is below that for a person without such disease. We can rarely say that some specific level is pathological in isolation from the larger context.

The twentieth-century concept of disease as a deviation from statistical physiological norms puts medical science center stage both in defining and in treating disease. This construction of the pathological as deviation from normal physiology has had three consequences. First, evil as a cause of disease has all but been removed from the picture;[5] no longer is disease associated with malign forces, but rather with measurable variations of normal phenomena; and diagnosis and treatment are now in the realm of scientific objectivity. Second, the presence of disease is determined by the medical profession, relying on measurements not readily available to the general population. It is no longer enough for a person to feel unwell to claim a disease; that status can be conferred only by quantitative measurements performed by qualified personnel.[6] A final consequence of this quantitative approach is that the goals of treatment are clear; treating the disease involves returning the deviant pathological parameters to normal physiological levels. This approach reinforces the centrality of the professional and the laboratory in the diagnosis and treatment of disease. Symptoms and bodily experiences of the affected individual are viewed as actually or potentially superfluous.

Given this quantitative conception of disease,[7] the role of medicine is to restore the normal state, to return the individual's physiological parameters toward the mean. However, even if we do accept statistical norms as the basis for some diseases, for other bodily parameters the statistical norm will not define the healthy (King 1981, 110). The ideal human being of medical textbooks is a 70-kilogram male with thirty-two teeth, no mental disorders, and a clean genetic slate. This ideal being is not statistically common, yet it colors our notions of health, and deviation from some of these ideal parameters constitutes disease. An overt and striking example of this is provided by R. A. Wilson (1966), whose ideal state of femininity includes a certain level of estrogen, such that women with lower levels suffer from an estrogen deficiency disease (MacPherson 1993, 146).

In addition to statistical and ideal norms, a further important normative component of disease is our evaluation of certain states; we desire to be healthy, and we dislike, or disvalue, certain states that we then call "ill health" or "disease." This conception of disease has the potential to reposition the experiences of the individual at the center of the definition of disease, yet it is our culture that largely determines which particular bodily experiences count as symptoms. Such cultural evaluation of symptoms leads to an intertwining of the moral and the medical; many practices (or symptoms) that have been culturally defined as morally unacceptable are redefined as medical problems. For example, during the eighteenth and nineteenth centuries, masturbation was regarded as a serious disease with a number of significant sequelae including dyspepsia, epilepsy, vertigo, deafness, headache, loss of memory, irregular heartbeat, and general loss of health and strength, as well as insanity and consumption. Hospital records from this period in the United States indicate that people were not only hospitalized for masturbation, they even died of it. There were specific treatments for the disease and documented cures (Engelhardt 1981).

J. W. Barbre has demonstrated a similar conflation of moral and medical values in nineteenth-century approaches to menopause. Nineteenth-century menopause was said to be associated with a wide variety of ailments,[8] not least among them moral insanity, in which previously moral women became unprincipled, peevish, and even suffered from fits of temper or self-absorption. The detrimental effects of menopause were exacerbated by engaging in "morally questionable" acts such as reading novels, attending the theater, or dancing. The treatment for menopause focused on protecting the delicate and fragile nature of women by avoiding excitement and resting (Barbre 1993, 29). This transformation of essentially moral values into medical diseases provides a powerful method for controlling undesirable behaviors; once defined as sick rather than immoral, society has a mandate to prescribe treatments, theoretically based upon medical science rather than cultural values.

In summary, our descriptive understanding of disease is supplemented by assumptions, more or less covert, concerning statistical, ideal, and evaluative norms. Whether or not we define a set of symptoms, or physiological parameters, as disease depends upon the cultural milieu as much as upon medical science, both of which can be equally effective in silencing the voice of the person experiencing the symptoms. This ambiguity in the definition of disease raises questions about the role of clinical medicine: Is it "because therapeutics aims at this state as a good goal to obtain that it is called normal, or is it because the interested party, the sick man, [sic] considers it normal, that therapeutics aim at it?" (Canguilhem 1989, 126). This is a crucial question about the nature of medicine. Accepting statistical norms as the basis for determining the presence of health and disease has the potential to exclude the experiences of the patient, thus excluding an important part of our understanding of ill health. Should medicine be concerned with treating deviations from statistical or culturally defined ideal states, or should medicine be concerned with the disvalue with which certain states are experienced by individuals? If we accept the latter position (that medicine alone cannot and should not rely upon normative values derived from physiological parameters), the way is open for individuals, as they experience their lives, to define what is or is not normal for themselves.[9] The goal of medicine, then, is to restore to these persons their own version of normal, such that their bodies return to that automatic state of function that is taken for granted before disease strikes.

Menopause

There are many ways of looking at menopause, each of which entails different conclusions. A biological essentialist approach sees menopause as a loss of true femininity; a scientific reductive approach defines menopause as a dysfunctional state; whereas feminist interpretations valorize it as a natural transition (Zita 1993). Each conceptualization of menopause involves different underlying assumptions about science and women. Their analysis adds to our understanding of the interrelations between culture and science.[10] My discussion is located within a biomedical model, focusing upon an analysis of two definitions of menopause as a disease. Menopause may be a disease because there are deviant values of certain parameters (the quantitative view), or because the woman concerned feels herself to be changed in some way that she experiences as ill health (the qualitative view). These two conceptions entail quite different consequences for women who are menopausal and postmenopausal.

A qualitative understanding accepts that menopause is a disease because women experience the associated bodily changes of menopause as a state of ill health. Descriptions of the symptoms of menopause vary, but a commonly accepted set of symptoms is listed in a leaflet called "A Balanced Approach to the Menopause." This leaflet, designed for distribution to women by their general practitioners, makes the following claims about menopause:

> These changes may result in both short term and long term effects. A lack of estrogen may cause symptoms such as: hot flushes, sweats, muscular aches and pains, more frequent urination, and loss of lubrication and elasticity of the vagina—which may cause painful intercourse. Infection of the vagina and bladder may also become more common at this time. Imbalance between the two hormones—estrogen and progesterone—frequently causes irregular and unpredictable bleeding. Many of the emotional symptoms, such as confusion, loss of memory, mild depression, and general irritability may be associated with these hormonal changes.

The leaflet continues, listing osteoporosis and cardiovascular disease as long-term effects of the lack of estrogen and finishing with the claim that "medical research has shown that estrogen replacement therapy reduces the risk of both osteoporosis and coronary heart disease" (Upjohn 1991). This quotation is from literature designed for patients, but it does not differ significantly from the lists of menopausal symptoms and sequelae given to general practitioners at educational meetings.

However, there is a considerable body of research that questions both the contents of such lists and how frequently such symptoms may be attributable to menopause. The authors of a Norwegian study of 2,349 women found that hot flashes, excessive sweating, and vaginal dryness were the only findings clearly attributable to the development of menopause (Holte and Mikkelsen 1991). These symptoms are generally accepted to be truly menopausal, but there may be racial and cultural variants even among this small core of symptoms. N. E. Avis and colleagues (1993) found that Japanese women aged forty-five to fifty-five experience hot flashes at the same rate as the general population, with no increase in the rate of flashes among menopausal women, thus challenging that most central of menopausal symptoms.

What seems certain is that the frequently claimed psychological symptoms of menopause, such as depression, irritability, and mood swings, are unrelated to menopause per se (Nicol-Smith 1996). The Massachusetts Women's Health Study found that naturally occurring menopause appeared to have no major impact on health and behavior. Almost one-quarter of the postmenopausal women in that study had no experience at all of hot flashes or sweats, and almost 70 percent did not report being bothered by their vasomotor symptoms (Avis and McKinlay 1995). Mary Lou Logothetis found that women experienced low levels of distress associated with menopause, with many of the women attaching overwhelmingly positive meanings to the experience (1993, 128).

Other studies have supported this with findings suggesting that many of the so-called symptoms of menopause are related to personal characteristics or preexisting health status rather than to menopause (Avis and McKinlay 1991; Dennerstein et al. 1993 and 1994; Groeneveld et al. 1993). The authors of an Australian study of 2,001 urban, Australian-born women reported that their findings suggested that medical and societal views about menopausal symptoms are likely to be based on the experiences of a particular segment of the population only. Women utilizing treatment had a broad range of psychosomatic symptoms compared with those women who did not utilize treatment. Interestingly, specific menopausal symptoms such as hot flashes were not significant in contributing to these women's utilization of services

(Morse et al. 1994). Given this bias in the populations of women seeking treatment, the research based upon these populations produces findings that may not be applicable to all menopausal women, yet the symptoms of these nonrepresentative populations shape the popular image of the menopausal woman.

Clearly, menopause is an event that, for many women, is acceptable as a normal part of life, in that the associated changes and features of menopause are not sufficiently distressing for the woman to declare herself unwell or suffering from a disease, or to seek medical treatment. For a number of women, this is not the case, and they do seek treatment for distressing symptoms. However, the number of women seeking treatment for menopausal symptoms may well be confounded by those seeking treatment for symptoms that are wrongly attributed to menopause (Dennerstein et al. 1993). Thus a qualitative understanding of disease concludes that menopause is associated with relatively few symptoms, the frequency and severity of which are widely variable. The individual finds herself well or not according to her own symptoms and experiences, but the majority are certainly not "diseased" in this sense of the word.

Let us now consider the quantitative approach to disease, in which not only menopause, but the whole of a woman's life subsequent to menopause, is considered to be an estrogen deficiency disease. This view originated in the 1960s with the work of Wilson, who claimed that "the menopausal woman is not normal; she suffers from a deficiency disease with serious sequelae and needs treatment" (Wilson 1963). This view currently remains vigorous, as stated recently by the (then) vice president of the Australian Menopause Society, who claimed: "The oestrogen deficiency state at the menopause should be regarded just like diabetes or thyroid disease" (Wyeth Pharmaceuticals 1991).

To accept menopause as a statistically defined deficiency disease, we must find the levels of estrogen in menopausal and postmenopausal women to be below those of "normal" women. This is the claim of Wilson and his followers. However, this claim does not stand up to scrutiny: if we measure the estrogen levels of thousands of postmenopausal women, we will find those levels to be *normal for their age*. Obviously these levels will be lower than those of women who are menstruating regularly, just as the levels of estrogen are lower in prepubertal women, but we do not describe female children as estrogen deficient. Similarly, Wilson and those who claim estrogen deficiency should question this definition in women who are in their sixties, seventies, and eighties. For with any other medical parameter, normal levels must be derived from the populations in question. We do not demand of seventy-year-old people that they have a creatinine level or blood pressure that is normal by thirty-year-old standards. Why, then, should we demand that a seventy-year-old woman have the estrogen level of a thirty-year-old unless our purely statistical definition of disease covertly includes a normative ideal of youthful femininity?

Perhaps this distinction between disease process and normal age-related change is untenable and there are some physiological changes that are a universal part of normal aging that are also diseases. Indeed, one author claims that there are "changes which occur with advancing age which appear to be normal characteristics of the aging process and which would also qualify as diseases" (Rowe 1984). This author offers several examples such as cataract formation and the age-associated rise in systolic blood pressure, as well as menopause. But here we need to distinguish very carefully between being postmenopausal and the occurrence of various adverse events such as osteoporosis and fractured hips. If disease is simply a quantitative variant of normal physiological values, then postmenopausal women do not have an estrogen deficiency disease; by definition, the majority have normal estrogen values for their age. This is not to deny the well-known association between advancing years and osteoporosis in women; certainly being postmenopausal is a risk factor for osteoporosis and its sequelae, but this does not make osteoporosis an estrogen deficiency disease per se.[11] The causes of osteoporosis are recognized to be multiple and include physical inactivity, low calcium intake, and smoking (Rose 1992). More specifically, being postmenopausal is only one of many risk factors for hip fracture; the others include low body weight, impaired balance and gait, muscle weakness, Parkinson's disease, long hip axis length, psychotropic medications, inadequate physical activity, being a member of the white race, and being female (Cumming 1996).

It is thus incorrect, given the quantitative model of disease, to call the postmenopausal state an estrogen deficiency disease. Is this merely semantic haggling, or are there consequences for the deficiency disease approach? Some research suggests that the latter is the case in that the type and severity of menopausal symptoms actually experienced by women may be significantly predicted by women's expectations of such symptoms (Abraham et al. 1994; Avis and McKinlay 1995). K. A. Matthews had similar findings, remarking that "it does appear that women's expectations about aspects of the menopause affect their psychological experience during the menopause" (Matthews 1992, 5). The health effects of a universal acceptance of postmenopause as an estrogen deficiency disease may well be impossible to measure, but it is not unreasonable to think that such effects exist and may be significant. Furthermore, the medicalization involved in declaring all menopause to be a deficiency disease lays the groundwork for accepting HRT as the sole effective treatment, thus excluding the possibility of alternative solutions to both menopausal symptoms and osteoporosis (Dickson 1993).

To sum up, using a definition of disease based upon the symptoms experienced by individual women, menopause is sometimes a disease but more often not. On a quantitative understanding of disease, all postmenopausal women are by definition diseased, but as I have argued, this definition is faulty and leads to untenable and undesirable conclusions. However, the idea that all menopausal women are diseased is publicly promoted in both popular and medical literature. This widespread acceptance of menopause as disease implicitly endorses HRT as the cure. Not only will HRT treat the disease of menopause, but HRT will also prevent osteoporosis and hence save the community the cost of treating osteoporotic fractures. I examine the evidence for this claim in the following section.

Treating Menopause

A qualitative view of menopause as a disease acknowledges that some women are unwell due to menopause and, as such, may be helped medically. There is good evidence that treatment with HRT decreases the incidence and severity of hot flashes and other symptoms (Khaw 1992), and at the level of the individual, there seems good reason to offer treatment to affected women. This says nothing about the desirability of asymptomatic women taking HRT to treat their so-called estrogen deficiency disease, in particular to prevent osteoporotic fractures.

There is evidence that taking HRT decreases the rate of bone loss in menopausal women (Hilliard et al. 1994), but there is as yet no good evidence that decreased bone loss will translate into a decrease in the fracture rate. The research into this question is not expected to provide a clear-cut answer until the year 2001. At present, expert opinion is divided over the degree of risk reduction for hip fractures offered by long-term HRT.[12]

Given the multifactorial nature of osteoporosis and fractures, serious questions arise about both the ethics of promoting an unproven treatment and the desirability of relying on a single drug therapy for prevention. However, the issue I wish to pursue here is the portrayal of women as excessive financial burdens to the community owing to their fragile bones. If it is the case that HRT will decrease the rate of fractures, is it reasonable to expect women to take it *for this reason*? Let us first consider the desirability of a campaign promoting drug therapy for a widespread medical problem (fractured hips) and then look at precedents for targeting and treating one section of the population because its members would otherwise impose a significant burden on the community.

In general, there are two kinds of public health solutions for health problems: reducing the risk of disease or searching for effective therapy (Harper et al. 1994). HRT may prove to be effective therapy for decreasing the rate of fractures in older women, yet it might be undesirable for a number of reasons. First, as any reading of public health texts shows, there is good evidence that treatment rather than prevention is an inherently limited way of approaching health problems on a population scale. Medical treatment may be wonderful and at times extremely effective (witness the treatment of infections with antibiotics, or the use of anesthetic agents for surgery), but overall population approaches aimed at reducing the risk of disease, such as vaccination, immunization, or compulsory seat belt legislation, are significantly better at improving the health of populations. For example, the decrease in male death rates from coronary heart disease from 450 per 100,000 in 1972 to 250 per 100,000 in 1992 is attributed to the reduction in the prevalence of risk factors, rather than to advances in medical treatments (National goals, targets and strategies 1994).

Second, classifying a large population of women as diseased is not without social and cultural effects. The metaphors used to describe menopausal women are those of disease and abnormal degeneration, metaphors consistent with a medical view of women's bodies as abnormal and out of control (Martin 1997; Dickson 1993). Such bodies require constant surveillance in the form of smears, mammography or bone densitometry, and multiple medical interventions for ubiquitous biological processes. The widespread promotion of HRT can be seen as one more attempt at repression and control, one more construction of the female body as faulty, potentially damaged, expensive, and generally inferior.[13]

Next, there is the effect of myth on symptomatology; G. L. Dickson has noted that "the metaphors of the discourses of menopause research contribute to the experiences of menopause for women" (1993, 54). This effect occurs both for individual women (witness the documented relationship between believing menopause to be a problem and then experiencing those problems [Abraham et al. 1994; Matthews 1992]) and at the social level. Using the concept of the body politic, J. N. Zita (1993) has explored the possibility of reading the symptoms of illnesses as metaphor for troubling social conditions. The current body politic does not welcome its "aging" female citizens, unlike men in their "prime," who monopolize social power and capital. It is not surprising that there is widespread promotion of a drug that will allegedly rejuvenate them.

Arguments based upon myth or metaphor can be difficult to substantiate. Cost-benefit claims seem, on the contrary, to be based upon concrete facts. However, these kinds of argument are equally difficult to analyze. What do we include as costs and what as benefits? Are the costs of treatment offset by the savings produced, or are different groups benefiting? Universal prescription of HRT reaps fine profits for pharmaceutical companies, but these profits are not available to the community at large. The cost of treating hip fractures is somewhere between Aus\$240–389 million per year (see note 2). It has been estimated that the cost of HRT for all Australian postmenopausal women would be Aus\$545 million per year (Prince 1996). The cost to individual women of attending medical practitioners, receiving and filling prescriptions, and taking medication on a daily basis is almost impossible to calculate in a meaningful way. Furthermore, even if HRT turns out to be an effective way of reducing osteoporosis-related fractures, it may not be the most cost-effective way. Other approaches, for example, campaigns aimed at decreasing smoking in young women and increasing levels of physical activity in women of all ages, may be more cost-effective.

An alternative way of looking at risk reduction is to ascertain the number of people who would need to be treated for one year to prevent one hip fracture. The most optimistic estimate of this number in women followed to eighty years of age is 500 women; this is for the highest risk group. For the lowest risk group, the number is 6,850. This means that up to 228 women might need to be treated with HRT for thirty years to prevent one hip fracture, a figure that is well above the generally accepted level of 500 person years (Henry 1995).[14]

Finally, I wish to examine precedents for the widespread prescription of medication as prophylaxis against a medical problem labeled as costly to the community. It is hard to think of any; a daily aspirin has been promoted as beneficial for middle-aged men to protect against expensive cardiovascular disease, but here the discussion is in terms of benefits for the individual men, not for the community at large. Perhaps the only parallel is an understanding of fertility as disease, for which taking oral contraceptives is the treatment. But here individual women benefit by avoiding unwanted pregnancies. The benefits of taking HRT are less easy to define, even in this limited context. It has been calculated that if a fifty-year-old woman takes HRT for thirty years, her life expectancy will increase by about one month (Grady et al. 1992). Of course, this says nothing about potential improvements in her quality of life; however, any calculation of quality of life must include the cost of the daily ingestion of medication in one form or another and the possible side effects of such treatment.

Clearly, there is not a good case either scientifically, morally, or economically for promoting the universal prescription of HRT to prevent
the occurrence of osteoporotic fractures.[15]

Summary

This discussion of menopause has highlighted some of the assumptions underlying our commonly accepted definitions of disease. As
well as a descriptive component, "disease" includes a significant normative one. The underlying ideal and evaluative dimensions are rooted
in cultural beliefs about medicine and women's nature. According to a
qualitative understanding of disease, individuals experiencing symptoms seek relief from medical treatment. On this understanding, menopause is a disease (for some women) for which we have effective medical treatment.

However, the more dominant quantitative understanding of disease
defines the abnormal as a deviation from a normal physiological value
and can be diagnosed only by medically trained individuals, not patients. Despite the dominance of this view, this is not an appropriate
way of describing the postmenopausal state, and it does not provide a
mandate for the universal prescription of HRT. In particular, conflating
postmenopause, estrogen deficiency, and osteoporosis cannot be justified. The postmenopausal state is not an estrogen deficiency disease
and is only one of many risk factors for fractures of the hip. Campaigns, such as those aimed at Australian general practitioners in 1993,
calling for universal prescription of HRT on the grounds of decreasing
the incidence and cost of osteoporotic fractures must be regarded with
deep suspicion. Treatment, rather than prevention, is a flawed solution
to medical problems. The characterization of menopause as a disease
promotes and reinforces the medical construction of female bodies as
faulty and diseased. It also influences the ways that women experience
their menopause. The cost-benefit arguments for treatment are based
upon dubious estimates and ignore the significant nonfinancial costs
involved in long-term ingestion of HRT. Furthermore, risk-benefit analyses for preventing fractures indicate that an unacceptably high number of women must be treated to prevent a single fracture. Worse yet,
there is no evidence that HRT will actually decrease hip fractures. Finally, there are no parallels in public health where one sector of the
community is expected to take lifelong medication to save money for
the community at large.

In sum, women, like men, have a right to make treatment decisions
based on their own interests. They also deserve high-quality informa-

tion about possible alternative courses of action. At present, such information is significantly absent.

Notes

I would like to thank Jill Need for helpful discussions in the preparation of this essay and George Tallis for help in finding reference material. Laura Purdy and Anne Donchin have provided invaluable editorial assistance and critical review. I would also like to thank Matthew Millar for his comments.

1. The use of the term "estrogen deficiency disease" dates from the work of R. A. Wilson in the 1960s and continues to the present day. In August 1993 Dr. Joseph Goldzieher, Emeritus Professor of Obstetrics and Gynecology at the Baylor College of Medicine in Texas, was a keynote speaker at a number of sponsored lectures in Australia. During his lecture, he made the claim that menopause is a deficiency disease like pellagra, scurvy, or beriberi, with the clinical consequence of osteoporosis. He reported that the cost of osteoporosis in the United States was U.S.$27 million per day (now estimated to be over U.S.$10 billion per year). A publication sponsored by Wyeth Pharmaceuticals available at the same time quotes a health economist reporting on Australian figures, claiming that the total cost to Australian society of nontreatment of menopause was Aus$150–200 million per year (Wyeth Pharmaceuticals 1991). The costs incurred by nontreatment were also raised at the launch of a video on menopause. This launch coincided with publicity for Estrapak and was aimed at general practitioners, 7,000 of whom attended the launch at multiple venues around Australia. See, e.g., McRea (1983) or Voda (1993) for further discussions of menopause as deficiency disease.

2. The figure of Aus$389 million comes from Wark 1996. The figure of Aus$240 comes from the number of hip fractures per year (15,000) multiplied by the median cost per person (Aus$16,000, calculated on direct clinical and welfare costs [Wark 1996]). This discrepancy demonstrates the difficulty and unreliability of estimates of costs.

3. See, e.g., Boorse (1981) for an example of the view that medical practice rests upon a foundation of value-free medical theory.

4. See also Callahan (1993) and Komesaroff et al. (1997) for two recent collections of papers on philosophical and cultural interpretations of menopause, and Sherwin (1992) for a feminist analysis of the relationship between Western medicine and women.

5. One notable exception to this is the moral hysteria surrounding HIV/AIDS.

6. Abnormal cervical smears, hypertension, raised cholesterol levels, and osteoporosis, among others, are examples of diseases that are diagnosed in asymptomatic individuals by health professionals. (This is the practice of screening [Last 1988].)

7. Despite obvious exceptions (such as inborn errors of metabolism), this model of disease remains influential in much medical thinking. It is interesting

to note that such a standard textbook of pathology as *Robbins Pathological Basis of Disease* does not contain a definition of disease, but merely lists the four aspects of diseases that form the core of pathology: etiology, pathogenesis, morphological changes, and functional effects.

8. These included chronic ill health, debility, consumption, rheumatism, ulcerated legs, diabetes, urinary tract problems, hemorrhoids, gout, tooth decay, heart disease, shingles, diarrhea and constipation, deafness, and cancer (Barbre 1993, 29).

9. By accepting that medicine must respond to those who feel themselves to be diseased, I do not wish to deny the benefits that follow from a quantitative understanding of disease, for example the early detection and cure of asymptomatic cancer of the cervix. Both understandings of disease are necessary for the best possible practice of medicine.

10. See Dickson (1993) for a discussion of four menopausal paradigms (biomedical, sociocultural, feminist, and postmodern) or Martin (1997) for a critical review of the current scientific metaphors employed in discussions of menopause.

11. It is interesting to note that osteoporosis is another example of a disease defined by variation of a physiological mean. The World Health Organization's definition of osteoporosis is "a disease characterized by low bone mass and micro-architectural deterioration of bone tissue leading to an enhanced bone fragility and a consequent increase in fracture risk." This is quantified by a densitometric definition of osteoporosis as a bone mineral density 2.5 standard deviations below the mean *for a young adult* (WHO 1994). Using this definition, 40 percent of women aged seventy to seventy-nine years have osteoporosis (Seeman 1996).

12. This point was debated at the 1996 Australian National Consensus Conference on the Prevention and Management of Osteoporosis. The extent of the decrease in fracture risk with HRT may be less than previously estimated and may be as low as 15 percent risk reduction rather than the more commonly quoted figure of 50 percent (Henry 1995). The duration of treatment is also controversial. At seventy-five years of age, even women who had taken ten years of HRT did not have a greater bone mineral density than women who had never been treated with HRT (Kelly 1996).

13. See Rogers (1997) for an exploration of possible reasons for this medical and cultural view of women's bodies as essentially abject.

14. For comparison, approximately 500 person years of treatment with antihypertensive drugs are necessary to prevent a single stroke when treatment is being used for mild to moderate hypertension: this is the benchmark of an acceptable benefit in current medical/social thinking (Henry 1995).

15. It is important to note that I have considered only the case of osteoporosis in this chapter; another whole set of claims and evidence exist concerning the allegedly beneficial effects of HRT upon the cardiovascular system. See Hemminki and McPherson (1997) for a recent review of the medical literature on this topic.

References

Abraham, S., D. Llewellyn-Jones, and J. Perz, 1994. Changes in Australian women's perception of the menopause and menopausal symptoms before and after the climacteric. *Maturitas* 20 (2–3): 121–128.

Avis, N. E., and S. M. McKinlay, 1991. A longitudinal analysis of women's attitudes toward the menopause: Results from the Massachusetts Women's Health Study. *Maturitas* 13 (1): 65–79.

———. 1995. The Massachusetts Women's Health Study: An epidemiological investigation of the Menopause. *Journal of the American Medical Women's Association* 50 (2): 45–9, 63.

Avis, N. E., P. A. Kaufert, M. Lock, S. M. McKinlay, and K. Vass, 1993. The evolution of menopausal symptoms. *Balliere's Clinical Endocrinology and Metabolism* 7 (1): 30.

Barbre, J. W. 1993. Meno-boomers and moral guardians: An exploration of the cultural construction of the menopause. In J. Callahan, ed., *Menopause: A Midlife Passage*. Bloomington: Indiana University Press.

Boorse, C. 1981. On the distinction between disease and illness. In A. Caplan, H. T. Engelhardt, and J. McCartney, eds., *Concepts of Health and Disease: Interdisciplinary Perspectives*. Reading, MA: Addison-Wesley.

Callahan, J., ed. 1993. *Menopause: A Midlife Passage*. Bloomington: Indiana University Press.

Canguilhem, G. 1989. *The Normal and the Pathological*. Trans. C. R. Fawcett and R. S. Cohen. New York: Zone Books.

Cotran, R. S., V. Kumar, and S. L. Robbins. 1994. *Robbins Pathologic Basis of Disease*, 5th ed. Philadelphia: WB Saunders Co.

Cumming, B. 1996. Epidemiology of Osteoporosis in Australia. Paper presented at Australian National Consensus Conference, 1996: The Prevention and Management of Osteoporosis, Canberra, October 23–24.

Dennerstein, L., A. M. Smith, C. Morse, H. Burger, A. Green, J. Hopper, and M. Ryan. 1993. Menopausal symptoms in Australian women. *Medical Journal of Australia* 159 (4): 232–236.

Dennerstein, L., A. M. Smith, and C. Morse. 1994. Psychological well-being, mid-life, and the menopause. *Maturitas* 20 (1): 1–11.

Dickson, G. L. 1993. Metaphors of menopause: The metalanguage of menopause research. In J. Callahan, ed., *Menopause: A Midlife Passage*. Bloomington: Indiana University Press.

Engelhardt, H. T. 1981. The disease of masturbation: Values and the concept of disease. In A. Caplan, H. T. Engelhardt, and J. McCartney, eds., *Concepts of Health and Disease: Interdisciplinary Perspectives*. Reading, MA: Addison-Wesley.

Grady, D., S. M. Rubin, D. B. Petitti, C. S. Fox, D. Black, B. Ettinger, V. L. Ernster, and S. R. Cummings. 1992. Hormone therapy to prevent disease and prolong life in postmenopausal women. *Annals of Internal Medicine* 117 (12): 1016–1033.

Groeneveld, F. P., F. P. Bareman, R. Barentsen, H. J. Dokter, A. C. Drogendijk,

218   *Wendy A. Rogers*

and A. W. Hoes. 1993. Relationships between attitude towards menopause, wellbeing, and medical attention among women aged 45–60 years. *Maturitas* 17 (2): 77–88.

Harper, A. C., C. D. J. Holman, and V. P. Dawes. 1994. *The Health of Populations: An Introduction,* 2nd ed. Melbourne: Churchill Livingstone.

Hemminki, E., and K. McPherson. 1997. Impact of postmenopausal hormone therapy on cardiovascular events and cancer: Pooled data from clinical trials. *British Medical Journal* 315: 149–153.

Henry, D. 1995. Meta-analysis of interventions for prevention and treatment of postmenopausal osteoporosis and fracture: Final report. Newcastle, Australia: University of Newcastle Osteoporosis Study Group.

Hilliard, T. C., S. J. Whitcroft, M. S. Marsh, M. C. Ellerington, B. Lees, M. I. Whitehead, and J. C. Stevenson. 1994. Long-term effects of transdermal and oral hormone replacement therapy on postmenopausal bone loss. *Osteoporosis International* 4 (6): 341–348.

Holte, A., and A. Mikkelsen, 1991. Psychosocial determinants of climacteric complaints. *Maturitas* 13 (3): 205–215.

Kelly, P. 1995. Estrogen therapy for osteoporosis. Paper presented at Australian National Consensus Conference, 1996: The Prevention and Management of Osteoporosis, Canberra, October 23–24.

Khaw, K. T. 1992. The menopause and hormone replacement therapy. *Postgraduate Medical Journal* 68: 615–623.

King, L. S. 1981. What is disease? In A. Caplan, H. T. Engelhardt, and J. McCartney, eds., *Concepts of Health and Disease: Interdisciplinary Perspectives.* Reading, MA: Addison-Wesley.

Komesaroff, P. A., P. Rothfield, and J. Daly. 1997. *Reinterpreting the Menopause: Cultural and Philosophical Issues.* New York: Routledge.

Last, J. M., ed. 1988. *A Dictionary of Epidemiology,* 2d ed. Oxford: Oxford University Press.

Logothetis, M. L. 1993. Disease or development: Women's perceptions of menopause and the need for hormone replacement therapy. In J. Callahan, ed., *Menopause: A Midlife Passage.* Bloomington: Indiana University Press.

Macpherson, K. I. 1993. The false promises of hormone replacement therapy and current dilemmas. In J. Callahan, ed., *Menopause: A Midlife Passage.* Bloomington: Indiana University Press.

Margolis, J. 1981. The concept of disease. In A. Caplan, H. T. Engelhardt, J. McCartney, eds., *Concepts of Health and Disease: Interdisciplinary Perspectives.* Reading, MA: Addison-Wesley.

Martin, E. 1997. The woman in the menopausal body. In P. A. Komesaroff, P. Rothfield, and J. Daly, eds., *Reinterpreting the Menopause: Cultural and Philosophical Issues.* New York: Routledge.

Matthews, K. A. 1992. Myths and realities of the menopause. *Psychosomatic Medicine* 54 (1): 1–9.

McRea, F. B. 1983. The politics of menopause: The "discovery" of a deficiency disease. *Social Problems* 31 (1): 111–123.

Morse, C. A., A. Smith, L. Dennerstein, A. Green, J. Hopper, and H. Burger. 1994. The treatment-seeking woman at menopause. *Maturitas* 18 (3): 161–173.

National goals, targets and strategies for improving cardiovascular health: Report of the National Health Goals and Targets Implementation Working Group on Cardiovascular Disease. 1994. Canberra: National Health Goals and Targets Implementation Working Group on Cardiovascular Disease.

Nicol-Smith, L. 1996. Causality, menopause, and depression: A critical review of the literature. *British Medical Journal* 313: 1229–1232.

Prince, R. 1996. Primary prevention of osteoporosis. Paper presented at Australian National Consensus Conference, 1996: The Prevention and Management of Osteoporosis, Canberra, October 23–24.

Rogers, W. A. 1997. Sources of abjection in Western responses to the menopause. P. A. Komesaroff, P. Rothfield, and J. Daly, eds., *Reinterpreting the Menopause: Cultural and Philosophical Issues.* New York: Routledge.

Rose, G. 1992. *The Strategy of Preventative Medicine.* Oxford: Oxford University Press.

Rowe, J. W. 1984. Physiological changes with age and their clinical relevance. In R. N. Butler and A. G. Bearn, eds., *The Aging Process: Therapeutic Implications.* New York: Raven Press.

Seeman, E. 1996. Biphosphonate therapy for osteoporosis. Paper presented at Australian National Consensus Conference, 1996: The Prevention and Management of Osteoporosis, Canberra, October 23–24.

Sherwin, S. 1992. *No Longer Patient.* Philadelphia: Temple University Press.

Upjohn Pty Ltd. 1991. A balanced approach to the menopause. Brochure. Rydalmere, New South Wales: Upjohn Pty Ltd.

Voda, A. M. 1993. A journey to the center of the cell: Understanding the physiology and endocrinology of menopause. In J. Callahan, ed., *Menopause: A Midlife Passage.* Bloomington: Indiana University Press.

Wark, J. D. 1996. Osteoporosis: The emerging epidemic. *Medical Journal of Australia* 164: 327–328.

WHO [World Health Organization] Study Group. 1994. Assessment of fracture risk and its application: Screening for postmenopausal osteoporosis. *World Health Organization Technical Reports Service* 843: 3, 5

Wilson, R. A., R. E. Brevetti, and T. A. Wilson. 1963. Specific procedures for the elimination of the menopause. *Western Journal of Surgical Obstetrics Gynecology* 7: 110.

Wilson, R. A. 1966. *Feminine Forever.* New York: M Evans.

Wyeth Pharmaceuticals. 1991. Anonymous articles in *Change of Life . . . Time for a Change.* Publication sponsored as a service to medical education, produced by the Medicine Group Pty Ltd, NSW, July.

Zita, J. N. 1993. Heresy in the female body: The rhetorics of menopause. In J. Callahan, ed., *Menopause: A Midlife Passage.* Bloomington: Indiana University Press.

Part III

Working for Change

12

Culture and Reproductive Health: Challenges for Feminist Philanthropy

Nikki Jones

Efforts to reach universal agreement on reproductive health ethics are complicated by pluralistic value systems and the problematic ethical nature of some reproductive health issues. Ruth Macklin recommends well-established ethical principles such as liberty, utility, and justice to create an ethical framework for reproductive health issues, but she warns that "basic principles may be applied differently by different societies, and changes in their application over time may be more or less vigorously resisted."[1] Feminists argue that the male-dominated field of bioethics needs a feminist perspective. Susan Sherwin identifies core themes in feminism that include

> a recognition that women are in a subordinate position in society, that oppression is a form of injustice and hence intolerable, that there are further forms of oppression in addition to gender oppression (and that there are women victimized by each of these forms of oppression), that it is possible to change society in ways that could eliminate oppression, and that it is a goal of feminism to pursue the changes necessary to accomplish this.[2]

Justice is required to remedy women's unequal access to resources and decision-making power that leads to many of their reproductive health problems,[3] and justice also shows that there is no moral justification for respecting certain existing situations;[4] but justice, by itself, does not tell us how to develop practical strategies to counteract the resistance of male-dominated cultures. This chapter examines the role of feminist philanthropy in enabling women to gain equal recognition of their reproductive health interests. It discusses how philanthropy can use its power to pursue gender equality, a form of social justice.

The Programme of Action of the 1994 International Conference on Population and Development (ICPD) is a significant document for those of us concerned with women's rights and reproductive health. For the first time, the reproductive health and rights of women and men (rather than the achievement of demographic targets) are central to an international agreement on population. The ICPD produced a definition of reproductive health[5] that was endorsed by the representatives of 184 countries, as well as detailed recommendations for action. The 1995 World Conference on Women confirms the ICPD Programme of Action and explicitly places sexuality within the context of human rights.[6] There is no legal obligation on signatory countries to implement the recommendations of these conferences, but they are part of a growing body of international customary law.

The late 1990s have seen unprecedented activity among UN and international organizations, multi- and bilateral donors, and, to a lesser extent, national government and nongovernmental agencies, as they seek to apply the Cairo and Beijing recommendations to their own policies and programs.[7] The resulting documents are impressive in that they represent (at least in writing) a move away from a single-minded focus on family planning to a holistic concept of reproductive health, framed in the language of the rights of couples and individuals.

While we acclaim the sentiments expressed in these documents, we might also reflect on the process by which they obtained international consensus. What is quite remarkable is that international standards have been set for many reproductive health and rights issues, apparently without qualifications.[8] Is this the triumph of international human rights over cultural relativism?[9] To what extent does this expression of Western democratic and individualistic values reflect a cultural hegemony on the part of donor and international agencies and conference delegates? Are these recommendations the expression of new and shared convictions among women and men worldwide or the result of skillful lobbying by Western feminist organizations?

I ask these questions not because I disagree with the recommendations, but because I am concerned with how they can be implemented in cultures with alternative values that accord women low status and little decision-making power and respond to claims about individual rights with counterclaims about community and family responsibilities. Can an approach be successful that ignores the power of local cultural beliefs and practices (or more precisely, the power of those men whose interests these cultures enshrine)? Can an approach be successful that fails to provide explicit opportunities to empower the oppressed? How can we create an enabling environment for reproductive health? What is an appropriate role for philanthropy?

Since the 1970s, we have witnessed international campaigns to promote child survival and safe motherhood. The reduction of infant mortality reflects wide agreement on strategies (including more readily available technological solutions)[10] and on the importance of the task. In contrast, the Safe Motherhood Initiative, launched in 1987 with the aim of reducing the estimated annual 500,000 maternal deaths by at least one-half by the end of the twentieth century, has a dismal record. Maternal deaths were in fact underestimated,[11] more than three million women have died since 1990, and the risk of dying as a result of pregnancy and childbirth fell by only 5 percent in five years.[12] This failure reflects a general reluctance to prioritize women's sexual and reproductive health needs and to improve their access to education and income. This failure is sometimes justified in terms of respecting local cultures, at other times in the language of structural adjustment. This suggests that if we are to improve women's lives, we need to understand and work with those who embody this resistance, as well as with those women and men who are the catalysts for change.

I am speaking as a British feminist[13] who has spent the last nine years working for a U.S.-based philanthropic organization whose mission is "to act as a resource for innovative people and institutions worldwide."[14] Since 1988, I have recommended funding to local organizations for the development and implementation of programs to improve the sexual and reproductive health of women and men: firstly in several West African countries[15] (based in Nigeria) and more recently in Southeast Asia, in the Philippines.

The West African countries where I worked are among the poorest on earth, with some of the highest levels of reproductive morbidity and mortality. It is estimated that pregnancy and childbirth cause the death of one in twenty-one African women and that, for every woman who dies, approximately thirty more suffer injuries and disabilities.[16] One-quarter of all women in the developing world have or will be affected by maternal injuries.[17]

Multiple factors are associated with high levels of maternal illness and death:

- an inhospitable physical environment with many pathogenic factors
- poverty, heavy burdens of work, and poor nutrition for women and girls
- early and continual pregnancy and childbearing

- lack of affordable, accessible, and appropriate health care
- local sexual, health, and birth practices
- women's lack of decision-making power and autonomy

Many of these conditions can be attributed to women's low economic and social status and to local cultural contexts. Here I focus on the role of culture, that set of beliefs, attitudes, practices, and power relationships that binds a group together and tends to be defined by the dominant group members.[18] As the latter are usually men, cultures often act to reinforce male beliefs, attitudes, and interests and to subordinate women's interests.[19] This does not mean that cultures do not change. The ideas, beliefs, values, practices, and relationships that define a culture are in constant evolution, subject to a variety of influences. Some of these are introduced from other (often Western) cultures, while some emerge from pressures and contradictions within the group itself. Increased expectations and access to decision-making power among the most disadvantaged members can be an impetus for cultural change. Women's empowerment can, perhaps, be measured by the extent to which they are able to define their own culture, one that reflects their own beliefs and sustains their own interests. Exposure to alternative resources and value systems can accelerate the demand for change among women and men.

When I arrived in Nigeria, I brought with me the power of money and the convictions and values of a Western feminist. I was aware of the dangers of imposing my own feminist perceptions and priorities on those with whom I was to work.[20] I also realized that the risks were much greater as a donor representative. It is common for donor agencies to determine the agenda, especially in a situation of scarce resources.[21] Even when priorities are not externally determined, a mere hint of donor interest in a particular topic can skew local research, action, and advocacy agendas, often controlled by men in search of funds. There is constant concern that donor agencies, armed with their own money and morality, can replicate the power relationships of colonialism in countries that still bear the imprint of many years of subjugation. A significant advantage for private philanthropies, being self-financed, is that they do not have to represent the interests of donor governments and thus have greater opportunities to develop policies and programs in response to local needs. Philanthropy also tends to be more explicit about its moral role. A major goal of my own institution is to reduce poverty and injustice. It was my responsibility in West Africa to ensure that these institutional goals could be effectively translated into improving the lives and reproductive health of women, the most disenfranchised group in these very poor countries. I could not ignore the considerable power that my position gave me. I felt that it was a moral responsibility as a feminist to recognize that power and to use it to promote women's interests. I therefore looked for ways my program could support local women in identifying and acting on their own reproductive health priorities.

I traveled widely; attended meetings; talked to donors, government officials, researchers, health workers, and activists; and listened to those whose voices are rarely heard, women themselves. It was Nigerian women who first introduced me to the common but neglected problem of vesico vaginal fistula (VVF).[22] The prolonged and obstructed labor associated with child marriage and early pregnancy, especially among poor and malnourished women, is a frequent cause of maternal and child death. Days of contractions "repeatedly grind down the skull of an already asphyxiated baby onto the soft tissues of a pelvis that is just too small."[23] Women who survive are often left with VVF and a dead child. Shortly after labor, the woman begins to leak urine continuously. Childless and emitting a permanent foul odor, in a culture that puts a premium on fertility and personal cleanliness, the woman is abandoned by her husband and secluded from society. In northern Nigeria in the 1980s, desperate VVF survivors traveled great distances to live on the steps of hospitals, sometimes waiting years for repair. Forced into prostitution in order to survive, they were exposed to reproductive tract infections, HIV, and AIDS. These are the "lucky" women who escaped death.

In an attempt to understand how cultures contribute to maternal death and disability, including VVF, we supported community-based research that identified hundreds of cultural beliefs and practices related to pregnancy and childbirth. In just one local government area in eastern Nigeria,[24] interventions to deal with prolonged and obstructed labor included:

- feeding a woman mucus from a donkey's nostrils
- drinking water in which the husband's trousers had been washed or water mixed with mud, clay, or chopped earthworms
- heavy massage; applying intense pressure and heat to the abdomen; rubbing the abdomen with the sandal of a woman who walks very fast
- inserting objects such as a boiled egg into the vagina
- starvation of the woman
- manual manipulation of the fetus through the vagina
- use of rusty knives or razors to cut the anterior wall of the vagina ("*gishiri* cut")

All these practices were designed to help the woman and to facilitate her labor, even the *gishiri* cut, a traditional form of episiotomy. Unfortunately, they can sometimes harm or delay access to effective health care. Some beliefs also deter women from asking for appropriate help, by linking prolonged labor with adultery or by imposing a rule of silence during labor.[25] We attempted to feed research findings into education campaigns for male community and religious leaders and women themselves, and to implement training programs for traditional and modern health care providers. Social science research thus provided empirical evidence about the need to support reexamination of existing moral values and practices concerning child marriage and early pregnancy.

Despite an estimated 80,000 new VVF sufferers per year[26] and a cumulative total of up to one million worldwide, stigmatization and shame have spread a blanket of silence over VVF in many countries. This reflects a moral framework that defines the value of a woman by her fertility but believes that women cannot act responsibly about reproduction. Fertility is therefore the source of both a woman's status and her subordination. A common male argument for controlling women's sexual and reproductive behavior is that women are driven by strong sexual instincts, making it impossible for them to act responsibly. This belief underlies two practices discussed in this chapter: the marriage of girls at or before puberty and female genital surgery. Women are blamed for prolonged obstructed labor and VVF because husbands associate this with sexual infidelity, not the fact that they themselves have impregnated a CHILD.

In Nigeria, with the exception of a few outstanding physicians who have dedicated themselves to VVF repair,[27] the issue was confined to medical textbooks. Kelsey Harrison[28] and Margaret Murphy[29] were pioneers in northern Nigeria in the late 1970s in studying the link among VVF, early childbearing, and poor nutrition. However, the impact of early pregnancy on maternal health is obscured by the fact that most women who suffer prolonged and obstructed labor die before they reach the hospital. It is difficult to form an advocacy movement of dead women.

For four years I supported a range of programs that sought to link evidence of women's pregnancy-related illness and death to a reexamination of the underlying moral beliefs and that encouraged practices that would promote better reproductive health for women. Partners included women's and advocacy groups,[30] professional health associations,[31] media practitioners, local and national government officials, universities, religious organizations,[32] male religious and community leaders (including Muslim emirs), district and village heads, and traditional chiefs. VVF was a powerful mobilizing issue because it involved the death of babies as well as the severe impairment of the woman's reproductive role, both highly valued. The challenge was to make the woman's own death or disability as important an issue as that of her fertility. A multisectoral National Task Force on VVF coordinated information gathering and dissemination, public education campaigns, training of health workers, community mobilization around preventive primary health care, and advocacy on women's status and age at marriage. Most Task Force activities were supported through grants to women's organizations that controlled the funds: this provided a strong impetus for male-led organizations to collaborate with women. Defying many setbacks over its six years of existence, the Task Force has played a unique role in raising awareness of and commitment to VVF prevention and management at both national and international levels.[33]

In Southeast Asia, despite an estimated maternal mortality rate eight times lower in the Philippines than in Nigeria,[34] community-based research in rural areas reveals that women still die from the same causes and that culture plays a critical role. Just recently, a team of local researchers in a remote mountain area in the southern Philippines found a woman who had died in obstructed labor and whose eyes and mouth had been sewn up to prevent the fetus emerging as an evil spirit. The researchers, women from the next village, fled in terror of being possessed themselves.[35] Such research is important because, unlike in West Africa, there is a general tendency to gloss over sometimes appalling pregnancy and birth conditions.

Of all Filipino women, 70 percent still give birth at home, half of them with only an untrained birth attendant to help them.[36] Many women are forced into unsafe abortions, leading to disability and death. The major cultural institutions in the Philippines, the Catholic Church and the family, stifle open discussion of abortion, which is referred to only in moral terms of sin and punishment. Research proving that abortion is a common event in women's lives has finally encouraged women health activists, researchers, and health professionals to take the first tentative steps toward advocating for the recognition of unsafe abortion as a public health issue. I am more than happy to support them in their advocacy, on the clear understanding that this is their own and not my agenda. For me to choose to prioritize work on abortion in this cultural context would be viewed as Western feminist cultural imperialism.

During my time in West Africa, I was frequently asked if I was working to eradicate female genital mutilation (FGM), or female genital surgery (FGS), as I prefer to call it.[37] After leaving Nigeria to live in Southeast Asia, I was intrigued when FGM was introduced into casual conversation by people who had never visited Africa. They seemed confused as to exactly what the term covered but decried it as the most barbaric of practices.[38] I was unaware at the time that these were merely faint echoes of a new furor about FGM in feminist circles and in the popular media following a book on the subject by the novelist Alice Walker.[39] The debate focused on whether we should be more tolerant and understanding of cultural practices other than our own, or whether they should be dismissed out of hand as savage torture. Views tended to be polarized. While Aman, a young Somali woman who had herself been circumcised, claimed that "they did it because they love me," Rosenthal, an American columnist writing for the *New York Times*, called this "the most widespread existing violation of human rights in the world."[40] I would argue that the disagreement is not about whether the practice of FGS does physical harm to women, but about the most ethical and effective ways of confronting this practice.

The whole debate around FGS raises moral questions for feminists. First, why have Western feminists chosen to highlight this particular practice? The sensationalist assertion that FGS is the most abhorrent of all traditional practices against women diminishes attention to other practices that not only injure but kill women. How much is this influenced by the importance attached in the Western feminist movement to the physical expression of women's sexuality? Surely it is our responsibility as feminists to help women in other cultures to define their own priorities and not to impose our own. The second moral problem is that some feminists are ready to inflict unequivocal judgments and solutions on women in other societies without sufficient understanding of the dynamics of local cultures.[41] Proposed responses are frequently linked to legislation to outlaw the practice and punish the "guilty" (usually women). I would suggest that this is both unethical, in that it assigns blame to women for a practice that is determined in a male-dominated culture, and often ineffective.[42] The very use of the terms "mutilation" and "torture" alienates these women, our most important partners in change.[43] The feminist movement has taught us that we can achieve real change only through a process of becoming self-aware and self-affirming. It seems to follow that the most valuable contribution we can make to the emancipation of women in other cultures is not to impose our own moral judgments (based upon our own cultural experience), but to enable them to make the changes they want within their own cultural context.

In Nigeria, I supported women's health organizations that had decided to address FGS in terms of its harmful impact on women's reproductive health.[44] These programs fully involved the practitioners in the awareness and education process. It was women from within the culture, often themselves survivors of the practice, who defined FGS as an issue and worked toward its eradication. Other women were more concerned with the cultural significance of FGS as a necessary initiation into womanhood.[45] In recognition of the importance of ritual and initiation in women's lives, some African women have made successful attempts to find ways to celebrate these rites of passage without genital surgery.[46]

The challenge, working for a U.S.-based philanthropy, is to find strategies that empower rather than impose change. Providing direct funding to local women's organizations and supporting affirmative action in favor of women in male-led organizations have substantially strengthened the voices of women. But it is also crucial to work with male religious and cultural leaders to create greater understanding and a more supportive environment for women's emancipation. We can always find, among both women and men, visionaries who are committed to a more equal and just society and who can work within their own religion and culture to achieve change.

Support for local participatory research is an important first step, as it highlights the impact of cultural beliefs and practices on women's health and well-being, and helps to mobilize women and men around these issues. Philanthropy can also provide opportunities to open up public dialogue about this new knowledge and its moral implications, providing fora where religious and cultural values that reinforce gender blindness and inequality can be discussed and challenged. Based on the recommendations that emerge from these exchanges, philanthropy can also provide the resources to support the actual implementation of change.

As a result of these experiences, I can suggest some strategies that might enable philanthropy to use its funding power to promote women's empowerment in reproductive health:

- Remember that we cannot define or solve the problems of women in other cultures.
- Support women within the local culture to identify and achieve their own goals.
- Make every effort to understand the local culture and belief systems and the opportunities and constraints that these present for women.
- Appreciate the role and meaning of religious and cultural rituals and encourage those not harmful to women and children.
- Look for unlikely partners.

- Support the voices of women and men with vision and commitment to change.
- Work with those who hold the power and define the culture (mostly men) in order to create space for women.
- Accept that strategies that reinforce religious and community leaders as forerunners of change are more effective than apportioning blame.
- Encourage multisectoral partnerships and understanding and respect among partners.
- Ensure that the language of advocacy is understandable and acceptable to all of those involved in the process of change.
- Link all efforts to improve sexual and reproductive health to efforts to improve women's education, income, and legal and social status.
- Allow time for change.

These are just some ways that I, as a feminist in philanthropy, have negotiated the thorny path toward reproductive health in countries where women's well-being is a low priority. This experience has convinced me that philanthropy can help empower women in other cultures by investing in their ideas, not in imposing its own.

Notes

1. Ruth Macklin, "Liberty, Utility and Justice: An Ethical Approach to Unwanted Pregnancy," *International Journal of Gynecology and Obstetrics*. Suppl. 3 (1989): 37.
2. Susan Sherwin, "Feminist and Medical Ethics: Two Different Approaches to Contextual Ethics," in *Feminist Perspectives in Medical Ethics*, Helen Bequaert Holmes and Laura M. Purdy, eds. (Bloomington, Indiana University Press, 1992), 29 n6.
3. Ruth Macklin, "Women's Health: An Ethical Perspective," *Journal of Law, Medicine & Ethics* 211, no. 1 (Spring 1993): 23, argues that an "ethical perspective on women's health begins and ends with principles of justice."
4. Macklin ("Liberty," 38) points out that "to conclude that a state of affairs ought to continue in the present and future because it existed in the past is a philosophical error."
5. *Programme of Action of the United Nations International Conference on Population and Development*, Cairo, September 1994. Chapter 7, "Reproductive Rights and Reproductive Health," para. 7: "Reproductive health is a state of complete physical, mental and social well-being and not merely the absence of disease or infirmity, in all matters relating to the reproductive system and to its functions and processes. Reproductive health therefore implies that people are able to have a satisfying and safe sex life and that they have the capability

to reproduce and the freedom to decide if, when and how often to do so. Implicit in this last condition are the right of men and women to be informed and to have access to safe, effective, affordable and acceptable methods of family planning of their choice, as well as other methods of their choice for regulation of fertility which are not against the law, and the right of access to appropriate health-care services that will enable women to go safely through pregnancy and childbirth and provide couples with the best chance of having a healthy infant. In line with the above definition of reproductive health, reproductive health care is defined as the constellation of methods, techniques and services that contribute to reproductive health and well-being by preventing and solving reproductive health problems. It also includes sexual health, the purpose of which is the enhancement of life and personal relations, and not merely counseling and care related to reproduction and sexually transmitted diseases."

6. *United Nations Report of the Fourth World Conference on Women*, Beijing, China, September 4–15, 1995. "Beijing Declaration and Platform of Action," para. 96: "The human rights of women include their right to control over and to decide freely and responsibly in all matters related to their sexuality, including sexual and reproductive health, free of coercion, discrimination and violence. Equal relationships between women and men in matters of sexual relations and reproduction, including full respect for the integrity of the person, require mutual respect, consent and shared responsibility for sexual behavior and its consequences."

7. The International Planned Parenthood Federation (IPPF), for example, has produced an *IPPF Charter in Sexual and Reproductive Rights* (London: IPPF, 1996) based on international human rights instruments and standards from four recent international UN conferences.

8. *Programme of Action of the ICPD*, sect. 8.25. Abortion was one of the issues seen as subject to laws prevailing in a particular country: "Any measures or changes related to abortion within the health system can only be determined at the national or local level according to the national legislative process."

9. Hilary Charlesworth, "Feminist Critiques of International Law," in *Third World Legal Studies, 1994–95*. (International Third World Legal Studies Association and Valparaiso University School of Law, n.d.), 9–10, examines the arguments of essentialism versus relativism within the context of feminist theory.

10. Oral rehydration therapy, immunization, and iodization programs have played a crucial role in reducing infant and child mortality.

11. WHO *[World Health Organization] Safe Motherhood* Issue 19 (1995): 3. Overall, maternal deaths were underestimated by 20 percent. In eastern, middle, and western Africa, earlier calculations appear to have underestimated by between one-quarter to one-third.

12. *Safe Motherhood Partners: Emphasizing Action* (New York: Family Care International, 1992). Given the increase in the number of women of childbearing age, the number of women who die each year has actually increased.

13. My definition of "feminist" is based on Rosemarie Tong, *Feminine and Feminist Ethics* (Belmont, CA: Wadsworth Publishing, 1993), 6: "A feminist con-

sciousness is political not only in the sense that it sees that women are subordinated (repressed, oppressed, suppressed) but also in the sense that it seeks to
eliminate this subordination." She identifies several feminist schools of
thought, including liberal, Marxist, radical, psychoanalytic, socialist, and postmodern feminism. I would describe myself as a socialist feminist in agreement
with Juliet Mitchell's argument that we must work on all of those structures
that overdetermine women's condition, production, reproduction, sexuality,
and the socialization of children if women are to achieve equality. I also share
Mitchell's conviction that, unless there is a change in a woman's interior world,
and she herself is convinced of her own value, external changes will not liber-
ate her (Juliet Mitchell, *Women's Estate* [New York: Vintage Books, 1971] and
Psychoanalysis and Feminism [New York: Vintage Books, 1975], quoted in Tong,
Feminine and Feminist Ethics, 9). I also agree with the postmodern feminist argument that women's experience is divided by class, race, and culture. I do not
think that these standpoints are contradictory.

   14. Michael Hooker, "Moral Values and Private Philanthropy," *Social Policy
and Philosophy* 4 no. 2 (n.d.): 4, suggests that "philanthropy as an institution has
an obligation to achieve a level of integrity well above that of society in general.
The overriding purpose of the institution of philanthropy is to improve the
world in all its aspects, particularly those that pertain to values. This purpose
cannot be accomplished well if philanthropic agencies do not themselves exemplify the highest ideals and values." He also argues that foundations should
be "more open in sharing information about their philosophies, goals, programs and procedures . . . including their own misgivings" (5). Of the Ford
Foundation in particular, it has been suggested that the institution has in the
past exhibited a "conspiracy of optimism" that has silenced doubts (7).

   15. These countries are Nigeria, Senegal, The Gambia, Togo, Burkina Faso,
and Mali.

   16. Ann Ward, *Nigeria Maternity Report, 1991* (Anua, Akwa Ibom State, Nigeria: St. Luke's Hospital, 1991). This hospital-based study in eastern Nigeria records 50 maternal deaths for 3,186 women delivering 3,266 babies (i.e., 1.6 percent of births resulted in the death of the mother).

   17. UNICEF (United Nations International Children's Emergency Fund),
The Progress of Nations Report 1996 (New York: UNICEF, 1996).

   18. A. M. Kleinman, "Cultural Issues Affecting Investigation in Developing
Societies," paper presented at the Institute of Medicine, National Academy of
Sciences Workshop, "Clinical Investigations in Developing Countries," Bel-
lagio, Italy, 1979, 1. "Clinical investigations in developing societies must be understood as taking place within the particular context of practical, everyday beliefs, values and power relationships that constitute local cultural systems"
(quoted in N. A. Christakis "Ethics Are Local: Engaging Cross-Cultural Variation in the Ethics for Clinical Research," *Social Science and Medicine* 35, no. 9
[1992]: 1080).

   19. Charlesworth, "Feminist Critique," 11, argues that "we need to investigate the gender of the 'cultures' that relativism privileges. Relativism is typically concerned with dominant cultures in particular regions and these are,

among other things, usually constructed from male histories, traditions and experiences." She adds that "for the term feminism to have any meaning, it must extend beyond local concerns."

20. Isabelle R. Gunning, "Female Genital Surgeries and Multicultural Feminism: The Ties That Bind, the Differences That Distance" in *Third World Legal Studies, 1994–95* (International Third World Legal Studies Association and Valparaiso University School of Law, n.d.), 17–18, outlines three steps for a Western feminist in approaching "culturally challenging" practices. "The approach involves: 1) understanding one's own historical context; 2) appreciating how the 'other' might perceive you especially as regards the colonial/imperial heritage that western feminists have been bequeathed; and 3) recognizing the complexities of the life and circumstances of the 'other' woman in her particular context."

21. In Nigeria in 1988–1992, support to family planning programs far outstripped any other reproductive health funding, a priority determined in Washington, D.C., and London, rather than in Lagos.

22. Josephine Effah, "Culture and Reproductive Rights," in *Constitutional Rights Journal*, April–June 1995, 27. The Nigerian National Task Force on VVF estimates that 100,000 Nigerian women have VVF and that for every 1,000 women who deliver, 2 are likely to develop VVF.

23. UNICEF, *The Progress of Nations Report, 1996*. It is estimated that obstructed labor is the cause of 40,000 maternal deaths per year, often following days of labor.

24. *A Report of the Nigeria Chapter of the Inter-African Committee on Traditional Practices Affecting the Health of Women and Children* (Gongola State, Nigeria: Mayo Belwa Local Government Authority, 1991).

25. In Nigeria, prolonged and obstructed labor is frequently attributed to the woman having an adulterous affair (partly because of an incorrect link with sexually transmitted infections), and it is believed that the woman will not be relieved until she confesses the name of her partner. The custom of *kunya* dictates that young girls must hide their first pregnancy, never ask questions or complain during pregnancy and childbirth, and must pay no attention to the first child when it is born.

26. UNICEF, *Progress of Nations Report, 1996*.

27. Some of the dedicated medical doctors in Nigeria who relieved the suffering of VVF survivors through fistula repair are U. G. Lister, John B. Lawson, Kelsey A. Harrison, Ann Ward, and Kees Waaldijk. They were also active and inspiring partners in attempts to move toward prevention.

28. Kelsey A. Harrison, "Childbearing, Health, and Social Priorities: A Survey of 22,774 Consecutive Hospital Births in Zaria, Northern Nigeria," *British Journal of Obstetrics and Gynaecology* 92, Suppl. no. 5 (1985).

29. Margaret Murphy, "Social Consequences of Vesico-Vaginal Fistula in Northern Nigeria," *Journal of Biosocial Science* 13 (1981): 139–150. Murphy found that the vast majority of these women had been married before their first menstruation, that this condition was frequently associated with a first pregnancy, and that it rapidly led to the loss of family support.

30. These included the National Council of Women's Societies of Nigeria, Women in Nigeria, the Nigeria Chapter of the Inter-African Committee on Traditional Practices Affecting the Health of Women and Children, and the Women's Health Organization of Nigeria.

31. These included the Society of Obstetrics and Gynaecology of Nigeria, the National Association of Nigeria Nurses and Midwives, and the Nigerian Medical Students Association.

32. These included the Medical Missionaries of Mary and the Young Men's Christian Association of Nigeria.

33. *Cairo Hearing on Reproductive Health and Human Rights*, NGO Forum, ICPD, September 4, 1994. Hajiya Amina Sambo, the founding coordinator of the Task Force, testified on "The Impact of Culture and Tradition on Women's Health in the Form of VVF."

34. UNDP (United Nations Development Program), *Human Development Report, 1995* (New York: UNDP, 1996). UNDP estimates of maternal mortality rates per 100,000 live births from 1980 to 1992 are 100 for the Philippines and 800 for Nigeria. The *Philippines Safe Motherhood Survey 1993* (Manila: National Statistics Office, October 1994), suggests that Philippine maternal mortality is higher than the UNDP estimate, at 209 deaths per 100,000 live births.

35. Personal communication with Rosena Sanchez, research coordinator, Ateneo de Davao University.

36. *Philippines National Safe Motherhood Survey, 1993*: 69.8 percent of women give birth at home (53.8 percent in urban and 84.8 percent in rural areas). Over half of these (51.5 percent) have only a traditional birth attendant to attend them (33.7 percent in urban and 68.1 percent in rural areas).

37. Gunning, "Female Genital Surgeries," 17.

38. D. W. Kaplan, S. D. Lewis, and J. Hammer ("Is It Torture or Tradition?" *Newsweek*, January 10, 1994) give an inaccurate description of female circumcision/female genital mutilation, despite their claim that "the details aren't in dispute: a girl, sometimes as young as an infant, has all or part of her external genitalia removed. That can mean excision of the clitoris and the labia minora. Then the surgeon—who typically isn't a doctor—scrapes the sides of the labia majora and stitches together the vulva with thread or thorns, all while the girls are awake or held down. The purpose, dating to ancient Egypt: to ensure virginity and eliminate sexual sensation, and thereby make women manageable." The process that this article describes is in fact infibulation, one of the most extreme forms of genital surgery, and the purpose stated here is one seen from a Western viewpoint.

39. Alice Walker, *Possessing the Secret of Joy* (New York: Simon and Schuster, 1993).

40. Quoted by J. Flint, "Putting Rites to Wrongs," *New York Times*, May 22, 1994, 25.

41. Gunning, ("Female Genital Surgeries," 20), reacting to one of Rosenthal's editorials in the *New York Times*, comments that "to describe the surgeries as a system of torture as opposed to an ugly piece of a much larger, more complex cultural value system is more than a value judgment. It so magnifies and

amplifies one aspect of an organic and multiple layered system of organization that it denigrates other aspects that are positive." She also argues that, by focusing on Africa countries, "he accesses an ugly set of racialized representations, myths, that have been formulated in the American psyche since the birth of the nation when the founding fathers had to justify the enslavement of African people."

42. Karungari Kiragu, "Female Genital Mutilation: A Reproductive Health Concern," special supplement to Population Reports, *Meeting the Needs of Young Adults,* series J, no. 41, vol. 33, no. 3 (October 1995): 3. Kiragu points out that the British colonial government passed a law against FGS in 1946 but that the practice still persists.

43. Kiragu ("Female Genital Mutilation," 1) argues that the use of the term "mutilation" accurately describes the consequences of the procedure. In my view, it inadequately describes, and even distorts, its cultural signification.

44. The Nigeria Chapter of the Inter-African Committee on Traditional Practices Affecting the Health of Women and Children and the National Association of Nigeria Nurses and Midwives encouraged communities to mobilize against FGS, organized national information campaigns, and trained modern and traditional health providers.

45. Marcus Mabry and David Hecht ("Women Fighting for Their Rites," *Newsweek,* October 14, 1996) document the reaction of the Bundo all-women secret society to Dr. Olayinka Koso-Thomas's attempts to eradicate FGS in Sierra Leone.

46. *East African,* August 12–18, 1996. Maendeleo Ya Wanawake, a group of Kenyan women activists against FGS, organize initiation rituals that celebrate the passage into womanhood without surgery.

13

Strategies for Effective Transformation

Barbara Nicholas

W hen I was a full-time mother, economically dependent on my partner, living in a small rural town, and working to improve the position of women in my church, the analysis of what to do and where to focus my energies for transformation was, I found, reasonably clear-cut. I could be part of political agitations and analysis, social support, and community action. It was a position I enjoyed. There was a sense of community, some goals to achieve. As feminists, we knew we were on the margins, victims of patriarchal oppression, with justice on our side. And we did some good work.

But now I have stumbled into a different world. I have a title to my name, and I am employed in a powerful institution (a medical school and university),[1] training people to be part of a powerful profession. I am well paid, I have got used to wearing jackets to work, and I find myself arguing with professors and sitting on national panels and committees that have the power to make decisions that will have a great impact on the lives of others.

My position and my options for action have changed, and with that change, I find the need to reflect on the consequences of a change in place and in position. What now is my agenda as a feminist? To what extent are the analyses and strategies that were useful and effective in one context able to be carried over into another?

Strategies that are effective, sustainable, and personally health giving when one is on the margins may not seem so successful once one has gained a place in the very institutions and discourses that feminism has analyzed and found wanting. I believe that one of the challenges for a feminist bioethicist concerns how to move to the "center"—to lose one's marginal status, to claim the power that it gives—without selling out, losing one's critical edge, becoming absorbed, assimilated, or domesticated. As bioethicists, we may have changed our personal positions (for instance, economically and socially), but there are many people who continue to be adversely affected by sexism, racism, homophobia, economic exploitation, and so on. Unless our ways of working affect the reality of those so oppressed, then we are not being effective, however successfully we hold down our jobs, publish, or sit on public bodies.

In this chapter, I want to explore some of the tensions for feminist bioethics as we seek to transform a conservative male discourse into a site of social change. This transformation needs to take place at several levels and can be effected in a number of ways. Here I will discuss two approaches: theoretical strategies that can subvert and undermine the patriarchal and other abusive assumptions written into the frameworks that structure the conversations in bioethics, and practical strategies for transformation within the institutions with which we work. My reflections are grounded in the New Zealand context and in my employment in a medical school, but I suspect that many of the tensions I experience are similar to those found elsewhere.

Negotiation of Place for Bioethics

Bioethics has now gained a place in health and research in New Zealand. As in many countries, it has been the narratives of the marginalized that have provided much of the impetus to change powerful institutions. For example, in 1988 it became known that for more than twenty years, women with the precursor to cervical cancer—carcinoma in situ—were not receiving standard medical care, but were being monitored and watched to observe the natural progression of the disease (see Cartwright 1988; Coney 1988). The women involved were informed neither that they were part of research nor that they were not receiving standard treatment. They thought they were receiving optimum medical care and being "kept an eye on." The doctor involved was one of the most senior medical gynecologists in the country, training a whole generation of doctors. Finally made public through the work of two nonmedical women, Sandra Coney and Phillipa Bunckle (recently elected a member of Parliament), a Commission of Inquiry was created, and long and detailed public hearings were held. This exposed entrenched sexism within medicine and a lack of reliable quality review of medical research. Massive changes have since been enacted in New Zealand, including an increase in medical ethics teaching, a Code of Rights for those who use health and disability services, and improved review of medical and other research involving human subjects. There is widespread public acceptance of the need to discuss the ethics of resource allocation, organ donation, assisted human reproduction, and the like.

This scandal has greatly assisted bioethics to negotiate a place for itself in research and in health care, particularly in medicine, which remains the most socially powerful of the health professions. But it is a continually negotiated place. For instance, in medicine, we cannot, as bioethicists, assume our right to be there or that our agenda is totally accepted. We are not marginalized or excluded, but we cannot impose or control. We have no authority other than that which medicine chooses to give us and which can be withdrawn at any time. Those who are given the responsibility for ethics education in medical schools are not thereby part of the dominant discourse. Our backgrounds may be in philosophy, law, theology, or other health professions. Many of us are "outsiders within," admitted to the medical discourse but not strictly of it, more than guests but not quite "family." Like in-laws, bioethicists come from a different family system; we belong, but in a particular and peculiar way.

The consequence of this is that bioethics must negotiate its place in the curriculum and do so in terms that are acceptable to the medical discourse. Howard Brody has identified one strategy that ethicists have devised: "[T]heir intended audience was not the morally defective physician who needed reform; it was the morally well-intentioned physician who wanted to do the right thing but was confused as to what the right thing was" (Brody 1992, 39). In this way, bioethicists have avoided confrontation and direct engagement with a powerful discourse. We have promoted ourselves as useful to medicine, helping the profession to do its job better. Bioethicists have recognized that our continued presence at the bedside requires us to phrase arguments in acceptable ways, in terms that can be heard and accepted by the medical world.

Yet many of the changes in medical practice to which bioethicists have contributed require voices of protest and resistance, the public articulation of subversive narratives and alternative knowledges. Like feminism, bioethics has been driven by the voices of those on the margins, those most vulnerable to powerful discourses and institutions (McNeill 1993; Rothman 1991). Bioethicists working with and within dominant discourses and institutions walk a delicate line with these alternative voices that provide the authority for their work. If we identify too closely, we risk dismissal for being too radical or outrageous. If we lose sensitivity to these voices, bioethicists risk co-option into medical discourse and the loss of our ability to do more than provide justification for the power structures of medicine.

For feminists, this fine line is even thinner. Not only are we negotiating the place of bioethics within medicine, we are also negotiating the place of feminism within bioethics. Dominated by the male Western tradition, and led by philosophers who did much of their foundational thinking prior to the latest waves of the feminist movement, bioethics discourse does not have feminist concerns and analysis woven into its fabric. Too often, I find that feminist concerns about the effect of a practice on women (or on those of a particular socioeconomic class, etc.) are seen as questions that can be tacked on later, interesting sidelines, women's work. For instance, a scholar was investigating the possibility of studying at my place of work. Her field of interest was justice in health care. That was the area of expertise of one of my colleagues, but the assumption was made that, as she wanted to focus on justice in health *for women,* this was a woman's issue and, therefore, she should work with me.

So feminist bioethicists must negotiate the risk of marginalization within both discourses. Bioethicists must keep their place within medical discourse and retain access to critical conversations; and feminists must endeavor to centralize feminist concerns and analyses in bioethics discourse and to transform the very terms of the conversations. The challenge is to find ways to do this that achieve a number of things: avoid unproductive confrontation or unwinnable battles; maintain personal integrity; retain and maintain a sense of what is needed to transform the lives of women who do not share our privilege or access to power; and, most importantly, find effective ways to "insert" feminist concerns into the dominant discourses and institutions that function to define normality or reality.

Strategies for Change

The two sets of strategies that I identify here are theoretical and practical. This is, of course, an artificial distinction: practice informs and stimulates theory, and theory transforms practice. Nevertheless, it is one way of structuring the conversation.

Theoretical Strategies

The most significant strategy in feminist bioethics is its resistance to specialization and the narrowing, confining approach that this can generate.[2] This has, of course, been a feature of bioethics in general; conversations have always been informed by the range of disciplines from which participants have come. As well as health professionals, philosophers, theologians, and lawyers have been involved in bioethics since the early stages of development of the discipline. However, feminist writings in bioethics continue to be informed by writers from areas as diverse as moral development, cultural studies, economic and social theory, philosophy of science, anthropology, law, philosophy, environmental ethics, third world and development issues, and critical pedagogy, to name but a few. One need only look at a review article such as that by Leslie Bender (1993), or the bibliographies in recent feminist bioethics publications such as those by Rosemarie Tong (1997), Susan Sherwin (1992), and Susan M. Wolf (1996), to see the enormous range of material that is seen as relevant to feminist bioethics.

This openness to multiple perspectives and the contribution of different theoretical approaches enables feminist bioethics to retain the (re)learning of this wave of feminism—that women's lives are affected and structured by multiple interlocking discourses and social practices. A great deal of feminist work has focused on making visible the variety of women's lives. Where are the women in this social practice? How does this policy affect women? Which women are benefiting from this way of organizing things? And how does this way of talking about things (or that way of organizing the data or applying the information) structure the possibilities for women's lives? It has become apparent that "women" are not the same,[3] not one amorphous group, having only one identity. This has always been obvious to some women, but middle-class, educated, white feminists have perhaps been slow to recognize their own privilege and the impact of factors such as race, class, and sexuality on the lives of other women.[4]

An early dualistic understanding of patriarchy, male versus female, was initially attractive to some feminists, but attention to the many different stories of women makes such an approach difficult to sustain. It is not only gender that structures the possibilities for women.[5] Issues of class, economic and social policy, race relations, and international relationships, as well as their effects on patterns of trade and employment, all have their impact. (Think only of the international trade in organs, the Human Genome Diversity Project, or the debates on biotechnology around the ethics of genetically modified food.) No one theoretical approach is sufficient to illuminate the complexity of different women's social positioning, nor the ramifications of topics such as surrogacy, euthanasia, genetic testing, or disability. Multiple tools and analyses are needed, and we need to find ways of conversing across disciplinary boundaries.

Openness to multiple disciplines has also helped us recognize common patterns in various theoretical constructs. This, in turn, has led to a strengthening and enrichment of feminist theorizing. The most obvious pattern is the extent to which dominant discourses have paid little or no attention to women's experiences. For instance, the male body has been defined as normal in much medical research, and Western philosophy has paid little attention to the moral lives of women as they bear children, maintain homes and relationships, and care for the sick and frail. It has been essential for feminist bioethics to focus on reproduction and on relationships in the family and with the vulnerable. We have focused on embodied experiences, valued the body as well as the mind, emphasized relationships. In doing so we have informed and transformed theoretical constructs. The importance of gender to ethics of care, dependency, social roles, and power has become more apparent.[6]

The second central feature of feminist bioethics is attention to issues of power. Woven through feminist bioethics is recognition of the postmodern context within which we work. Multiple voices, multiple realities are now part of our social world. No one metanarrative (apart from the fact that there is no one metanarrative!) seems to be able to embrace the diversity of experiences or the many ways of making meaning.

A rejection of the possibility of an absolute position may lead to an open relativism, where no moral judgments can be made, where no values or perceptions can be seen as more valid than any other. If no rational grounds can be found to justify the particular positions, if no tradition can provide significant grounds to convince others to enter it, then each must walk her own path and allow others to follow theirs. This position takes an extreme form in liberal arguments for autonomy, such as found in H. Tristram Engelhardt's *Foundations of Bioethics* (1989). With no place to stand that can be universally justified, respect for the autonomy of others becomes the immovable commitment. The only ground for interference in another's worldview is if they encroach upon one's freedom by limiting one's autonomy.

The difficulty with such a "live and let live" approach is that it ignores the fundamental issues against which many discourses protest: the power of a dominant discourse to define reality and the parameters of a conversation. Engelhardt's approach, for instance, makes invisible or secondary/derivative concerns for justice and social mediation of power (through race or gender). He is able to argue that indentured labor is more acceptable than taxes, as one freely chooses to enter into a contract in the former. He pays no attention to the circumstances that may pressure one into "choosing" to sign such an arrangement. His approach does not require that inequalities within communities be addressed, as it is assumed that individuals have the freedom/resources to opt in or out of particular communities or traditions if they find the structuring of networks of relationship unacceptable. Such an assumption fails to take account of people's social positioning within communities that compromise their freedom. Nor does it recognize that even if opting out of a particular community is possible, there may not be an alternative community where factors such as race and sex are not significant mediators of one's social power or lack thereof.[7]

Relativism is a position that functions to support/protect those who are part of the dominant discourse. It fails to address the power of the dominant discourse to maintain social-political-economic structures, nor does it provide criteria for dealing with power imbalances between discourses or communities. If the dominant discourse or tradition sees some practice as acceptable, relativism protects it from external critique. As Elizabeth Grosz has argued, "Relativism amounts to an abdication of the right to judge or criticize a position—any position—and a disavowal of any politics insofar as all positions are rendered equivalent" (Grosz 1993, 194). Challenge is only possible if the marginalized discourse can find terms internal to the dominant discourse into which its concerns can be translated. An example of the difficulty of challenge to a discourse if one is from a different tradition would be female circumcision. Western culture finds the practice unacceptable; but with no source of authority on which to base a challenge, it has had difficulty finding arguments acceptable even to itself (Sherwin 1992, 61ff.). Recent arguments have rejected the practice on the basis of criteria internal to the communities that practice it (James 1994).

Feminists can find themselves coming close to this position as they embrace many of the arguments of postmodernism.[8] As we recognize and value difference, we run the risk of fragmenting and reconstructing the subject of any discourse, to the point where we no longer have a political place to stand. If all women are different, then on what does feminism stand? If we no longer have a common community of resistance, if we are no longer victims of the same patriarchal oppression (but possibly also members of social groups that perpetuate and benefit from other forms of oppression), then it can no longer be our common victimhood that drives and motivates our political work or informs our engagement with the work of social transformation (hooks 1984). Feminism is challenged by postmodern discourse to identify the moral basis of its resistance to the dominant discourse, and the use of the power we gain we find ourselves in socially powerful positions.

Sherwin offers an approach here. She argues for a position that acknowledges and respects relativism but that maintains a commitment to the wrongness of oppression:

What feminist ethics claims is that oppression is a pervasive and insidious moral wrong and that moral evaluation of practices must be sensitive to questions of oppression, no matter what other moral considerations are also of interest. Such analysis requires an understanding not only of the nature of oppression in general but also of the nature of specific forms of oppression. (Sherwin 1992, 57)

Sherwin also argues that one must pay attention to the place of power in establishing moral standards. She accepts, with the relativist, that we have access to nothing more foundational than community standards. There is no metaethical or metaphysical reality to which we can appeal as ultimate authority. But she argues that we must pay attention to the manner in which standards are reached by a community. The presence of moral standards per se does not mean those standards are worthy of acceptance. If the moral standards of a community are to be worthy of acceptance and of trust, then the process by which those standards are reached must itself be evaluated. If standards of the community are reached through coercion, ignorance, exploitation, or indifference, then they themselves can have no legitimate standing.

In judging whether or not the standards reached by that community are morally acceptable, she places emphasis on the position of those who are most vulnerable:

A feminist moral relativism demands that we consider who controls moral decision-making within a community and what effect that control has on the least privileged members of that community. Both at home and abroad, it gives us grounds to criticize the practices that a majority believes acceptable if those practices are a result of oppressive power differentials. It will not, however, always tell us precisely what is the morally right thing to do, because there is no single set of moral truths we can decipher. Feminist moral relativism remains absolutist on the question of the moral wrong of oppression but relativist on other moral matters. (Ibid., 75)

Recognition of the multifaceted forms of oppression and the variety of social-political positions of women, as well as a clearly articulated position on the moral wrong of oppression, provides feminist bioethics with a basis of resistance to a simplistic liberal stance that would appear to inform much bioethics discourse. There is a tendency in much that is written to reduce ethical issues to ones of autonomy and choice. (I think here, for example, of arguments in support of surrogacy, abortion, and incentives to participation in research.) Such recognition also provides a basis for resisting postmodern fragmentation of political action.

As bioethics continues to resist specialization, embrace diverse experiences, and attend to issues of power, different models, languages, and images have made their way into "male-stream" publications. We now have Tom Beauchamp and James Childress (1994) (in their almost canonical publication and approach to bioethics) discussing an ethics of care; narrative ethics (which makes room for the diversity of lived experience) has credibility; and feminist bioethics is making its way into traditional journals.[9] This is an important beginning. But the task continues to transform the terms of the conversation within "male-stream" bioethics. What I think we are seeking to do is to reshape the old question in ethics: How should we live? It is not, How shall we live in a world in which we are normal, define reality, hold power—but how shall we live in a world where inequality, oppression, and exploitation are a daily reality in people's lives and define the context within which they must make decisions about health care, relationships with the environment, uptake of new technologies, care of the frail, and reproduction?[10]

Practical/Political Strategies

However stunning our theoretical sophistication, there are certain practical realities in working for social transformation within institutions and discourses that have social power. I find health professionals as individuals to be caring people, committed to doing their job well, but the terms of the discourses have been laid down in a patriarchal (and racist, homophobic, etc.) society. As we engage with these people and the discourses in which they (and we) are fluent, I think there are three tasks upon which we need to focus.

First, we need to find ways to survive in institutions that may have no commitment to social change. We need to establish a variety of forms of community that will both care for us and keep us accountable to these women who do not share our privilege and for whom we work. We can do this through various forms of networking.

We need to work with other women committed to social change. They may not be in our institution or immediate context, but if we are to sustain feminist work, we need a sense of community—a recognition that we are not alone, that other women also are working to make the world a different place. In male-dominated institutions like medicine and universities, many feminists are quite isolated in their workplaces.

We further need to establish networks and credibility with other colleagues who may or may not share our feminist analysis. Moving from the margins to the center can mean that we see the world in oppositional terms: them versus us. And nonfeminists can often see feminists in the same oppositional way. But the discourses of the center are sites of contest and debate. There are many voices there with whom we can cooperate and from whom we can learn. It is important that we invest time and energy in establishing common ground and healthy, respectful relationships with others.

And we need to maintain our contacts with other discourses and experiences to stay in touch with the grass roots of our community, with those who are frequently silenced or ignored by powerful institutions. This is an area where I find an enormous challenge. The emotional and psychological impact of marginalization becomes harder to remember as I become surrounded in a cocoon of financial security, education, and access to privilege. We need to hear what is happening to people in different social situations, their pain and dilemmas and experience. We need to be responsive, and in some senses accountable, to their lives. If we cannot hear the experience of the marginalized, then moving to the center has done nothing more than swap a few people around without changing anything fundamental.

All these strategies for survival feed our souls. It can be soul destroying to work in institutional contexts that can see male as normal, that assume the appropriateness of hierarchy, and where the critical questions of feminism are rarely validated or welcomed. The people may be (and frequently are) wonderful individuals, but I often feel quite alien, as though I am from another planet. Now repositioned into a dominant institution, I find the need to be reminded, recalled to analyses and frameworks of thought that make meaning or sense to those excluded from power and that provide permission to ask the unimaginable questions.

Second, we need to find ways to change the culture. It is not sufficient to insert ourselves into the dominant institution if all that happens is that we become "one of the boys." This changes nothing other than giving a few privileged women access to well-paying jobs and more status. We need to find ways to change the management culture of our institutions so that they can move away from hierarchical and often abusive ways of working. How can we encourage young women and other underrepresented groups? How can we transform the informal mentoring practices that often function to reproduce privilege? What changes do we need to make to publishing patterns, to appointment procedures for committees and public bodies?

The purpose of these changes, in my mind, is to do more than promote an alternative group to power. The goal is to fundamentally change the ways in which an institution and a culture function, to systematically transform privilege and status so that it will benefit those on the margins more than those at the center. This is a difficult orientation to maintain. The more we receive the rewards of participation in patriarchy (tenure, promotion, status, wealth, and influence), the more we stand to lose if we change the system. It can, at times, be a difficult process to discern what is appropriate self-care; when is it appropriate to protect ourselves from being battered by the institutional contexts or to retain a place where we can effect change? And when are we allowing ourselves to be domesticated and no longer a disturbing presence?

And third, we need to maintain hope. This is not always easy. Changes struggled over for years disappear overnight; we educate our colleagues, then they move on and other men take their places and we must begin all over again! I find it tempting to limit my horizons, to be grateful that I have a job, to talk about "being realistic" about what can be achieved. But such resignation is a form of privileged despair, one I can afford only because I am not anywhere near the margins. Given how many are still marginalized and disadvantaged in health care and in society, it is important that we find ways to keep alive our sense that many things are possible, as well as necessary. We need to find ways to celebrate the changes that have taken place, to honor the small shifts as well as the big ones. We need to share the stories of strategies that have worked, of people who have changed their research practices or their interactions with patients, of institutions that have changed the ways they offer services or have revamped their education process. We need to hear one another's stories so that we know more is possible, just as we need to hear stories of those nearer the margins so that we know that more is needed (Welch 1990).

Conclusion

Feminist bioethics is at an interesting phase in time. In Western bioethics, the recognition of a need for a gender analysis of issues is becoming more widespread, and women are becoming more visible in health care and in academic institutions. This is an important beginning, but it is only a beginning.

Women's movements, along with other movements for social change and justice, have effected some important gains. A number of concerns, as well as a number of individual women, have moved from margins to center. But a large number remain on the margins. We need to continue to reflect on our strategies, at both practical and theoretical levels, to ensure that we are being effective not only in continuing to bring more of the margins into the center, but also transforming the center itself.

Notes

1. The Bioethics Research Centre of Otago University, New Zealand.

2. For a discussion of the destructive and confining effect of specialization, see Macedo 1994. Although about education, not bioethics, he argues that the fragmentation of knowledge that specialization encourages leads to an "inability to make linkages between bodies of knowledge and the social and political realities that generate them" (22).

3. For discussions of essentialism in feminist thinking, see Fuss 1989; Hirsh and Keller 1990; DiQuinzio 1993; and Spelman 1988.

4. For examples of writings in ethics that emphasize these issues, see Cannon 1988; Hoagland 1988; and Welch 1990.

5. And as women's lives have become more visible (in all their variety), it has also become apparent that the experience of some men has also been rendered invisible by the assumptions that all men [sic] are equal. Yet only some men live lives of privilege and status. Many others experience the downside of inequalities due to their social, economic, and political positions.

6. For instance, there is an extensive literature that has emerged from Carol Gilligan's work (1983), which has influenced both feminist and nursing theory and hence has had an impact on bioethics. See, for example, Bender 1993; Carse 1991; Cooper 1991; Davis 1992; Hekman 1995; Larrabee 1993; Parker 1990; Tronto 1993; and Uden et al. 1992.

7. As Stanley Hauerwas points out in reference to Engelhardt's work and the concept of freedom for which he argues, "freedom too often turns out to be but a name for the power that some exercise unjustly against others" (Hauerwas 1988, 13).

8. For discussions about the relationship between feminism and postmodernism, see DiQuinzio 1993; Flax 1990; Fuss 1989; Hirsh and Keller 1990; Nicholson 1990 and 1992; and Spelman 1988.

9. See, for example, the Spring 1996 edition of the *Journal of Clinical Ethics*.

10. See Bell 1993, Cannon 1988, Hoagland 1988, and Wendell 1990 for examples of explorations of what it means to have moral agency in an imperfect world, where power is often abused and so often written into dominant and pervasive discourses.

References

Beauchamp, T. L. and J. F. Childress. 1994. *Principles of Biomedical Ethics*. 4th ed. New York: Oxford University Press.

Bell, Linda. 1993. *Rethinking Ethics in the Midst of Violence*. Lanham, MD: Rowman & Littlefield.

Bender, Leslie. 1993. "Teaching Feminist Perspectives on Health Care Ethics and Law: A Review Essay." *Cincinnati Law Review* 61: 1251–1276.

Brody, Howard. 1992. *The Healer's Power*. New Haven, CT: Yale University Press.

Cannon, Katie. 1988. *Black Womanist Ethics*. Atlanta, GA: Scholars Press.

Carse, Alisa. 1991. "The Voice of Care: Implications for Bioethical Education." *Journal of Medicine and Philosophy* 16: 5–28.

Cartwright, Sylvia. 1988. *Committee of Inquiry into Allegations Concerning the Treatment of Cervical Cancer at National Women's Hospital and into Other Related Matters*. Auckland: Government Printing Office.

Coney, S. 1988. *The Unfortunate Experiment*. Auckland: Women's Health Action.

Cooper, Mary Carolyn. 1991. "Principle-Orientated Ethics and the Ethic of Care: A Creative Tension." *Advances in Nursing Sciences* 14 (2): 22–31.

Davis, Kathy. 1992. "Towards a Feminist Rhetoric: The Gilligan Debate Revisited." *Women's Studies International Forum* 15 (2): 219–231.

DiQuinzio, Patricia. 1993. "Exclusion and Essentialism in Feminist Theory: The Problem of Mothering." *Hypatia* 8 (3): 1–20.

Englehardt, H. Tristram Jr. 1989. *Foundations of Bioethics*. New York and London: Oxford University Press.

Flax, Jane. 1990. *Thinking Fragments*. Berkeley, CA: University of California.

Fuss, Diana. 1989. *Essentially Speaking: Feminism, Nature, and Differences*. New York: Routledge.

Gilligan, Carol. 1983. *In a Different Voice*. Cambridge MA: Harvard University Press.

Grosz, Elizabeth. 1993. "Bodies and Knowledges: Feminism and the Crisis of Reason." In *Feminist Epistemologies*, ed. Linda Alcoff and Elizabeth Potter. New York: Routledge.

Hauerwas, Stanley. 1988. *Suffering Presence*. Edinburgh: Trent Clark.

Hekman, Susan J. 1995. *Moral Voices, Moral Selves: Carol Gilligan and Feminist Moral Theory*. Cambridge: Polity Press.

Hirsh, Marianne and Evelyn Fox Keller, eds. 1990. *Conflicts in Feminism*. New York: Routledge.

Hoagland, Sarah Lucia. 1988. *Lesbian Ethics*. Palo Alto, CA: Institute of Lesbian Studies.

hooks, bell. 1984. *Feminist Theory: From Margin to Center*. Boston: South End Press.

James, Stephen. 1994. "Reconciling International Human Rights and Cultural Relativism: The Case of Female Circumcision." *Bioethics* 8 (1): 1–26.

Larrabee, Mary Jeanne, ed. 1993. *An Ethic of Care*. New York: Routledge.

Macedo, Donaldo. 1994. *Literacies of Power*. Boulder, CO: Westview Press.

McNeill, Paul M. 1993. *The Ethics and Politics of Human Experimentation*. Cambridge: Cambridge University Press.

Nicholson, Linda J. 1990. *Feminism/Postmodernism*. New York: Routledge.

———. 1992. "On the Postmodern Barricades: Feminism, Politics and Theory," in *Postmoderism and Social Theory*, Steven Seidman and David Wagner, eds. Cambridge, Oxford: Blackwells.

Parker, Rand Spreen. 1990. "The Search for Relational Ethics of Care." *Advances in Nursing Sciences* 13 (1): 31–40.

Rothman, David. 1991. *Strangers at the Bedside*. New York: Basic Books.

Sherwin, Susan. 1992. *No Longer Patient*. Philadelphia: Temple University Press.

Spelman, Elizabeth. 1988. *Inessential Women: Problems of Exclusion in Feminist Thought*. Boston: Beacon Press.

Tong, Rosemarie. 1997. *Feminist Approaches to Bioethics: Theoretical Reflections and Practical Applications*. Boulder, CO: Westview Press.

Tronto, Joan. 1993. *Moral Boundaries*. New York: Routledge.

Uden, Giggi, Astrid Nurberg, Anders Lindsein, and Venke Marhaug. 1992. "Ethical Reasoning in Nurses and Physicians Stories about Care Episodes." *Journal of Advanced Nursing* 17: 1028–1034.

Welch, Sharon. 1990. *A Feminist Ethic of Risk*. Minneapolis: Fortress Press.

Wendell, Susan. 1990. "Oppression and Victimization." *Hypatia* 5 (3): 15–46.

Wolf, Susan M., ed. 1996. *Feminism and Bioethics: Beyond Reproduction*. New York and Oxford: Oxford University Press.

14

Women and Health Research: From Theory, to Practice, to Policy

Françoise Baylis, Jocelyn Downie, and Susan Sherwin

This chapter explores our experience as members of a feminist research group attempting to influence Canadian public policy on research ethics. It reviews some of our efforts to ensure that principles of feminist ethics be incorporated into new national ethics guidelines governing research involving humans. The experience has been fascinating, frustrating, challenging, sometimes encouraging, and sometimes depressing. We focus here on some of the frustrations we experienced in this process because we believe we now have a better understanding of these frustrations, and we hope our insights will be of value to other feminists who seek to shape public policy. We draw particular attention to the common themes that emerged between our theoretical and practical concerns in this process.

We are members of the Feminist Health Care Ethics Research Network (hereafter the Network), a multidisciplinary group of eleven researchers and practitioners with an interest in feminist ethics and women's health. The Network was supported from 1993 to 1997 by a grant from the Social Sciences and Humanities Research Council of Canada. Its mission was to reflect on the nature and implications of feminism for health care ethics.

The activities discussed here are the interactions between the Network and the Tri-Council Working Group on Research Involving Humans (hereafter the Working Group). The Working Group was created in the fall of 1994 by the three principal funding agencies controlling public research money in Canada: the Medical Research Council of Canada, the Social Sciences and Humanities Research Council of Canada, and the Natural Sciences and Engineering Council of Canada. The Working Group was assigned the task of drafting a common set of ethics guidelines for research involving humans, which could be adopted by all three councils.

Early in the process, the Working Group invited public input. The Network decided to accept this invitation and to engage actively in the policy-making process with a view to influencing the research guidelines from a feminist perspective. We determined that participation in the public debate was necessary for us to sharpen our own understanding of the issues in question and, on this basis, to promote morally appropriate policies and practices.

In this chapter, we do not review the details of the full history of our involvement with the Working Group,[1] but rather focus on a few of the many issues that we debated: (a) the direct exclusion and underrepresentation of women as subject-participants[2] in health research; (b) the indirect (but intentional) exclusion of some women from clinical research by requiring the use of hormonal contraceptives as an inclusion criterion; and (c) the failure to attend to women's interests in setting the research agenda. Through reflection on our interaction with the Working Group on each of these specific issues, we attempt to ground and illustrate our concerns about feminist participation in the policy-making process. We write in the hope that these reflections will assist others who are equally committed to the transposition of their theoretical work to the realm of practice.

Theory and Practice:
The Direct Exclusion and Underrepresentation of Women as Subject-Participants in Health Research

A major focus of our submissions to the Working Group concerned the problem of direct exclusion and underrepresentation of women in research. Research guidelines currently promulgated by the three councils are silent on the matter of gender equity. Specifically, they say nothing about the need to include women in research and, perhaps more importantly, nothing condemning the use of gender as an exclusion criterion.[3] Our goal was to persuade the Working Group to issue an unqualified condemnation of all health research that excluded women except where the disorder or disease under investigation occurred exclusively in men.

In our original written submission, we criticized the exclusion of (some) women from clinical studies and argued against the extrapolation of data from male subject-participants to female patients:

Historically, [research projects on] many diseases that are common to both sexes have systematically excluded women from participation; as a result, the data necessary for making treatment decisions for women are unavailable and must be inferred from data collected about men, even though there are important physiological differences between men and women that make such inferences problematic. Even when some data are collected about women's responses to the treatment in question, we may lack information about how specific groups of women will be affected (e.g., those who are disabled, elderly, pregnant, or poor). Women (and many other oppressed groups) ought to be represented proportionately to their health risk in any clinical studies likely to be of specific benefit to subject populations. (P. 1)

We insisted that the revised guidelines require researchers to provide an explicit ethical justification for any decision to limit research participation on the basis of gender (and any other traits associated with oppression such as race, class, ethnicity, age, religion, disability, and sexual orientation) so as to ensure that the implicit social devaluation of the oppressed group would not be a factor in decisions about research participation.

Prior to the completion and circulation of the first official draft of the new research guidelines, an early draft of the relevant section of the guidelines was shared with us by an individual member of the Working Group. The draft text was encouraging; it stated:

Guideline: The funding bodies should engage in educational efforts that will ensure that investigators are aware of gender biases and that studies are equitably conceived and designed with respect to gender.
Explanation: Research must employ study designs with gender distribution by age, risk factor, incidence/prevalence, etc., appropriate to the research objectives.

Guideline: The funding bodies should study the attitudes and institutional barriers to participation in research among women, racial and ethnic groups, and the poor.
Explanation: Even when included in study design women may experience barriers to participation. These barriers—some beliefs and some very practical, especially for women living in poverty—must be addressed if recruitment is to be truly inclusive and fair.

Guideline: Investigators should tailor study design and recruitment and retention issues with an understanding of the methodological and ethical issues involved in research with particular populations.
Explanation: "One size does not fit all" in research when sensitivity to issues of exclusion and oppression is at stake. (Pp. 4, 5)

Input on this unofficial draft text was requested, and, adhering to the format dictated by the Working Group, we suggested the following re-wording. Our intention was to restate the original points more force-fully:

> *Guideline:* The research subject/participant population should be as repre-sentative of the patient population as possible. It follows that women (and members of other under-researched populations) must be included in clinical trials in adequate number. Mere inclusion, however, is not suffi-cient. There must also be valid gender (and other) subgroup analysis of the research data; otherwise women (and members of other under-re-searched populations) may be included in the research and yet remain in-visible. Also, as appropriate, the research design must consider variability created by hormonal fluctuations.
> *Explanation:* The human body is differentiated by gender, race/ethnicity, class, culture and poverty. Valid subgroup analysis of the research data is necessary, therefore, to identify possible differences between different populations as regards both efficacy and side-effects. The requirement that hormonal fluctuations be considered in the trial design is based on recent evidence that the phase of the hormonal cycle may significantly ef-fect factors under study.

> *Guideline:* Research that excludes women as research subjects/partici-pants or ensures their under-representation should not receive REB [Re-search Ethics Board] approval and should not be funded unless the inves-tigators provide an ethically compelling justification of the decision to exclude, or limit the participation of, women (e.g., research focused on an attribute that pertains only to men).
> *Explanation:* It is important that women not be unjustly excluded from, or under-represented in, research. Research data on drug dosages, device ef-fects, treatment regimens and side-effects obtained from male-only trials may not be generalizable to women.

> *Guideline:* The recommendation that women be included in research is not an invitation to exploit women as research subjects/participants. Re-search that exploits women is unacceptable.
> *Explanation:* It is important that women (especially disadvantaged women) not be exploited by research that exposes them to risk without any potential benefit.

To our surprise and dismay, the first official draft *Code of Conduct for Research Involving Human Subjects* reduced all of the above proposals to a single prescriptive clause that clearly sought to mask the gendered nature of the problem of exclusion and underrepresentation:

When it is possible and appropriate, the researcher must endeavour to recruit women as well as men as research subjects. Researchers must provide the REB with an ethically compelling justification for the decision to
exclude, or limit the participation of, women or men in a research protocol. (Pp. 12–13)

When the opportunity arose, again as a result of unofficial consultation
by a(nother) member of the Working Group responsible for redrafting
this section of the guidelines, we suggested that the text be revised.
Most importantly, we stressed the need to remove the clause "when
possible and appropriate." In our view, this clause created an unacceptably broad exception to a general rule of inclusion. We also emphasized the importance of requiring appropriate subgroup analysis of the
data, not merely the inclusion of more women subject-participants. For
political reasons, however, we decided not to insist on representation
proportionate to health risk (as in our original submission), but instead
to make the simpler claim for equal representation. Also for political
reasons, we decided not to challenge the reference to men and women
as though the problem of exclusion were common to both sexes. While
we strongly objected to this effort to mask the problem of gender ineq-
uity, we thought the more important goal of including women in research might be best achieved by not tackling this issue. The statement
about inclusion of men wouldn't diminish the claim about including
women, whereas challenging the claim about men risked angering
those unwilling to acknowledge that women had ever been inappro-
priately excluded from clinical research. In consideration of the above,
we proposed the following revision:

Researchers must recruit women and men as research participants in
equal number and, as appropriate, must undertake relevant subgroup
analyses of the research data. This is important both for the validity of the
research and social justice in the distribution of the benefits of research
and research participation. Researchers who intend to exclude, or to limit
the participation of, women or men must provide the REB with an ethically compelling justification for the decision to do so.

This suggested revision was not adopted, and the text from the first
official draft reappeared almost verbatim in the next interim draft of
the guidelines, now titled *Code of Ethical Conduct for Research Involving
Humans*. In fact, the only changes between the two drafts (which at first
glance might appear as minor wording changes) were in the final sen-

tence and had the net effect of further weakening the strength of the claim—the claim that researchers must provide the REB with an ethically compelling justification for the decision to exclude research participants on the basis of gender was softened such that researchers should (but presumably needn't) provide the REB with an ethically acceptable justification for their decision.

The interim draft of the guidelines was not released publicly, but a member of the Network (whom we will call F.B.) was invited (though not in this capacity) to be one of five external reviewers to comment on the draft at a closed meeting of the Working Group. At this meeting, F.B. stressed the inadequacy of the clause about women's participation in research. She also noted that the previous prescriptive clause concerning research participation by members of racial, ethnic, and cultural groups did not limit the researcher's responsibility to "when possible and appropriate." Why, then, was this caveat in place for research involving women? F.B. also emphasized the fact that, despite all of the Network's efforts, there seemed to be little (if any) understanding of the need for subgroup analysis of the research data. From the beginning, we had not merely advocated that researchers include women subject-participants, but that they undertake appropriate subgroup analysis of the data, so that relevant differences between men and women might be identified and further investigated. Somehow, this critical point had been omitted from both official drafts of the guidelines.

The only other relevant prescriptive clause in the interim draft of the guidelines stated that "presumably fertile women and those who are pregnant or breastfeeding should not be automatically excluded as research participants" (p. 8–3). It was pointed out to the Working Group that following this clause, infertile women could be automatically excluded from research. Why was there resistance to including a simple statement to the effect that "no women should be automatically excluded from research on the basis of gender, including women who are pregnant"? In this way, one could make a strong statement regarding the inclusion of women in research and also emphasize that this claim applied equally to all women, including those who are fertile, prepubescent, postmenopausal, celibate, lesbian, and so on.

After more than two years of "engagement" in the policy process, we could not understand the Working Group's continued resistance to the inclusion of a strong statement guaranteeing women subject-participants equal access to the benefits of research. This was particularly perplexing in light of the fact that the parallel requirement for inclusion on racial and ethnic grounds was suitably strong (i.e., it did not include the qualifying clause "where possible and appropriate"). Our reflections on this led us to ask the following questions: Did we, unwittingly, harm the interests of women by drawing attention to this issue, thereby galvanizing those opposed to gender inclusion to carefully limit any such inclusion requirement? Was there more fear of a strong gender inclusion requirement than a strong race inclusion requirement because women have been very vocal and organized around this issue and so might be more forceful in their efforts to enforce the requirement? Or was it simply that the voices of researchers were more powerful than those of women subject-participants, such that their interests were subordinate to those of the researchers?

To our surprise, the final version of the *Code* prepared by the Working Group explicitly acknowledged that "the inclusion of women in research is essential if men and women are to equally benefit from research." (p. 6–3) There was also a strong prohibition against the unwarranted exclusion of women. Article 6.4 stated: "No women should be automatically excluded from relevant research" (p. 6–4).

We were certainly pleased with these statements; however, we had serious reservations about the explanatory text that followed, and we also had serious objections to the previous article—Article 6.3:

> Researchers and REBs must endeavour to distribute equitably the potential benefits of research. Accordingly, depending on the themes and objectives of the research, researchers and REBs must:
> (a) select and recruit women from disadvantaged social, ethnic, racial and mentally or physically disabled groups; and
> (b) ensure that the design of the research reflects appropriately the participation of this group. (P. 6–4)

In our view, the phrase "depending on the themes and objectives" seriously undermined the force of the article. Further, the article failed to recognize that it is not only recruitment, but also retention, that must be addressed if there is to be a serious commitment to the equitable distribution of the benefits of research. In addition, the article failed to recognize that it is not only recruitment and retention that matters, but also appropriate subgroup analysis of the data.

In the final document released by the councils—the *Tri-Council Policy Statement on Ethical Conduct for Research Involving Humans*—the two articles cited above are combined in Article 5.3, and the claim limiting the scope of the article (i.e., "depending upon the themes and objectives") remains.

> No woman should be automatically excluded from relevant research. Researchers and REBs shall endeavour to distribute equitably the potential

benefits of research. Depending on the themes and objectives of the research, researchers and REBs shall:

(a) apply equitable criteria in selecting and recruiting women from disadvantaged social, ethnic, racial and mentally or physically disabled groups; and

(b) ensure that the design of the research reflects appropriately the participation of this group. (P. 57)

This change conflates the concerns about the exclusion of women in general, the exclusion of multiply disadvantaged women, and the exclusion of pregnant and breast-feeding women in particular. These forms of exclusion raise different concerns that require, but are not given, different types of responses. Furthermore, a caveat is introduced explaining that the phrase "depending upon the themes and objectives" means that the article is "relevant to some but not all research" (p. 56). This caveat clearly further undermines the force of the article.

Contraception as a Mandatory Inclusion Criterion for Women in Clinical Research

In our review of the first official draft of the revised guidelines, we were shocked to learn that these guidelines permitted researchers to design research protocols in which contraceptive use by women could be a legitimate inclusion criterion. The permission, while not stated in such explicit terms, was evident in the expressed tolerance for this sort of practice. The guidelines merely warned researchers to be aware that they could not "take for granted a shared view on that issue [i.e., contraception] should they insist that all women participating in a particular study be on contraceptives" (p. 12–5). In our written response to the first draft, we objected strenuously to this position:

> It suggests that researchers are at liberty to insist that women research subjects be on contraceptives. This leaves women of childbearing potential at risk of exclusion from research should they refuse to take contraceptives. The issue is whether the women are at risk of becoming pregnant, not whether they are on contraceptives. The text, as phrased, improperly suggests that all women of childbearing potential are heterosexually active. (Pp. 6, 7)

Later, at a public meeting to discuss the first draft, one member of the Network (F. B.) raised this issue again and pointed out to members of the Working Group in the audience that here was evidence of unacceptable gender bias, given that no equivalent contraceptive criterion applied to men. The Working Group was reminded of the fact that prospective women subject-participants are capable of making appropriate informed choices for themselves, such that the only requirement needed in the guidelines was one concerning an obligation to explain the relevant research risks and the options available to avoid or end pregnancy while participating in research (e.g., abstinence, barrier contraception, termination of pregnancy).

We thought the point straightforward and uncontroversial, and, in reviewing the relevant section of the interim official draft of the research guidelines, we believed that we had had a sympathetic hearing. The Working Group explicitly acknowledged that "excluding women or requiring contraception in women but not men created an ethically unjustifiable imbalance" (p. 8–3). However, on closer inspection, we found that the problematic sentence on contraceptive use had simply been moved elsewhere in the guidelines (p. 3–8).

We again objected to this offensive clause and pointed out that, taken together with the clause that women need only be included in research "when possible and appropriate," there was effectively no assurance that the inappropriate exclusion and underrepresentation of women in research would be effectively addressed. Any researcher could insist on contraceptive use as a condition of research participation and subsequently argue that a refusal to use hormonal contraceptives was an "ethically acceptable" reason for not including women subject-participants. In other words, the wording continued to allow researchers an escape clause for excluding women from research, and it allowed researchers who did include women to treat them as incapable of making responsible decisions about their own reproductive well-being.

In marked contrast, the final version of the *Code* prepared by the Working Group critiqued the requirement of contraceptive use for research involving women of childbearing potential. A further positive change was the explicit acknowledgment of the fact that

> many women have been automatically excluded from research (e.g., the possibility of pregnancy was used as justification for excluding presumably fertile women, especially those not using contraceptives). Concerns for the embryo, fetus or a new-born infant were used as justifications for excluding pregnant or breast-feeding women. Excluding women, or requiring contraception in women but not men, created an ethically unjustifiable imbalance (see Section I). (P. 6–4)

What pleased us, however, did not please the three councils—the *Tri-Council Policy Statement* does not include the above text. There is no direct reference to "the exclusion of women of child-bearing age from drug trials because of possible harms to potential offspring" (p. 6 -1).

F. Baylis, J. Downie, & S. Sherwin

There is also no recognition of the fact that no similar exclusion applies to men of reproductive age (p. 6–2). Instead, the exclusion of women of childbearing age from drug trials is summarily described as a form of indirect (and inadvertent) exclusion:

> Exclusion from research has also arisen indirectly. Concerns about legal liability associated with particular populations have prompted the exclusion of women of child-bearing age from drug trials because of possible harms to potential offspring, for example. (P. 55)

This limitation aside, it is explained that Article 5.3, which stipulates that women "should not be automatically excluded from relevant research," is

> clear about presumptive or automatic exclusion from research on the basis of gender or reproductive capacity. If in the past many women have been automatically excluded from research on such grounds, Article 5.3 rejects such an approach as a discriminating and unethical use of inclusion or exclusion criteria. (P. 57)

Though, in principle, we object to the conditional nature of the claim regarding past exclusion (i.e., "If in the past many women have been automatically excluded . . ."), we nonetheless believe that a remarkable shift has occurred that should better protect and promote the interests of women.

Setting the Agenda: Determining the Research Priorities

Throughout our involvement with the Working Group, we not only argued that women (and members of other oppressed groups) should be included in research as subject-participants, but also that women should be included as researchers and active participants in the deliberations regarding research priorities. In our view, it is critical that more women be included in the processes by which research decisions are made and carried out, their underrepresentation to date having resulted in a research agenda that is largely unresponsive, if not contrary, to the interests of women and other oppressed groups. Our original written submission to the Working Group insisted on this point:

> While the research agenda regarding women's health needs has historically neglected many important questions, there has been a substantial body of research directed at gaining control over women's reproduction. In this area, women have received a disproportionately large share of research attention, and, as a result, women must now assume an unfair

share of the burden, risks, expenses, and responsibility for managing fertility because that is where the knowledge base is. . . . Further a very large share of the entire health research budget is absorbed by clinical studies directed at conditions that threaten those who are most privileged in society. For example, even though the links between poverty and illness are well known, health research often focuses on ways of responding to illness rather than avoiding it in the first place. (P. 2)

To better ensure that the research agenda was sensitive to the needs and interests of women, we argued that the process of setting the research agenda and conducting the research needed to be rethought, possibly along the lines of a partnership model:

In order to fulfil the moral obligation to eradicate oppression, we must begin by challenging the process by which research agendas are set and research programs are carried out. We might, for example, need to [consider] . . . an alternate conception in which research is pursued as a collegial activity; under this model, subjects and investigators collectively negotiate the terms of participation and the goals of the activity. (P. 3)

In reviewing portions of an unofficial draft of the guidelines prior to the release of the first official draft, we were encouraged to see some recognition of the need for shared power in determining research priorities. This objective was translated as follows in that interim document:

Guideline: Efforts must continue to encourage women of all racial and ethnic groups to engage in research and to participate in those bodies which set and support the research agenda.
Explanation: The inclusion of women in research populations is essential if all at risk are to benefit from the findings of the study. Exclusion raises serious concerns regarding generalizability of the data. (Pp. 4, 5)

Pleased as we were to see our concern partially reflected in this draft, we thought the proposed claim was not strong enough. Within the style constraints stipulated by the Working Group, we suggested an alternate wording, which, in our view, stated the point more persuasively:

Guideline: Funding Agencies (private and governmental) and REBs should work to remove institutional barriers to women's participation in research. In addition, they should actively encourage and enable women to become research subjects/participants, researchers, members of REBs, and members of those organizations that set and support the research

agenda. In so doing, there should be particular attention to the need to include women of colour, women who are members of cultural or religious minorities, and women who are poor.
Explanation: The research agenda may be unresponsive to the interests of women. One way to address this concern is to ensure that women, and not just women in the white middle classes, are recruited as research subjects/participants, and encouraged to be active researchers, REB members, and members of organizations that determine the research priorities.

Surprisingly, the first draft of the *Code* did not include either version of the wording. Indeed, it completely failed to discuss the importance of ensuring appropriate representation by and of women in the organizations that set the research priorities. To be precise, Sections 2, 3, and 4 of the *Code*, which discussed the responsibilities of research institutions, sponsors of research, and the granting councils, did not discuss the context in which research is conducted. Moreover, there was absolutely no mention of the need to ensure appropriate participation by women in all aspects of the research endeavour. Silence on this issue was unexpected since the Working Group member responsible for drafting this section of the document had seemed sympathetic to our concern, as indicated by the original wording.

In our response to this draft, we reiterated our concern and again recommended to the Working Group that it attend to the issue of underrepresentation not only with respect to participation in research protocols, but also with respect to participation in the processes by which research priorities and projects are chosen and pursued. Further, aware that the Working Group might continue to resist discussing the ethics of who sets the research agenda and who determines what counts as research, we asked—in the alternative—that a disclaimer be included in the introduction to the revised guidelines acknowledging the fact that these guidelines did not address all ethical considerations associated with research involving humans.

In assessing the interim draft *Code* for the external reviewers' meeting, F.B. noted that this draft also failed to include any discussion of who sets the research agenda. It also neglected to include any kind of disclaimer. The omission was especially troubling since there seemed to be an obvious place for the matter to be addressed: the guidelines identified five stages of research, the first of these being the planning stage. But what of the preceding stage in which the research priorities are determined? From our perspective, this omission seemed not only erroneous, but naive. How could the Working Group fail to appreciate that the planning stage of research is *very much* influenced by a prior stage of the research endeavor in which priorities are set regarding the research to be promoted and funded?

In a last attempt to influence the final document, F.B. argued, yet again, that the first stage of research involves the setting of the research agenda (by the councils and other granting agencies, as well as by governments, by journal editors, and by academic and professional bodies) and that the broad agenda directly affects (limits) the planning stage of research. Resorting to argument by authority, F.B. cited the National Forum on Health report in which it is explicitly recognized that the enhancement of research that benefits Canadian women must include efforts at defining a research agenda that is gender sensitive and inclusive.[4] In response, the Working Group continued to insist that this issue was beyond its mandate.

An important question for us became why we were failing to have this very important question of principle acknowledged. There were several possible explanations, and perhaps more than one applies. First, we questioned the strategy we adopted in our effort to advance this particular concern. When we first identified this issue, we thought its importance self-evident. We could not see how new guidelines for research involving humans could be drafted responsibly without attending to the fact that it matters who sets the research agenda and how, since this determines the nature of the research to be promoted and funded. When told by the Working Group that this issue was not part of its mandate, we argued otherwise, believing that if our arguments were persuasive, the Working Group would simply choose to broaden its mandate.

Discouraged by the response of the Working Group, we wondered whether we should have spent more time trying to understand the nature of the resistance we were encountering and considered other ways of responding to that resistance. It did not seem to us that our arguments were weak, but rather that the Working Group did not feel empowered to address the issue. Reiterating moral arguments seemed to fail to provide the needed momentum. Did we need, then, to step back from the political process that we were actively engaged in and think about the composition and agenda of the Working Group itself? In doing so, might we have observed that the very issue we were concerned about was being played out before our eyes? The Working Group's own research agenda had been set by external bodies, and it was ultimately answerable to them. The mandate it was assigned determined its research agenda. It was given the task of developing a set of guidelines that would be acceptable to those agencies, and it may well have felt the constraints such a task entails. To effect change in the mandate, it may have been necessary to lobby the sponsoring agencies and not merely the Working Group. After all, the councils set the agenda the Working Group was trying to conform to, just as we claimed is usually the case in clinical research.

Given the process that had gone before and the state of our own reflections on this issue, we were pleasantly surprised to find that in the final version of the *Code* submitted to the three councils, the following text appeared:

> There are multiple agents involved in setting "the research agenda", that is, the general direction of current and future research. . . . To be effective in setting and maintaining a just research agenda, there must be action and intelligent interactions by all those directing the research project. (P. 6–2)

From our perspective, there was little substance there, but we were willing to claim a minor victory for succeeding in having the issue named.

In the *Policy Statement* released by the councils, mention of the research agenda remains, but the text is substantially changed.

> Whether it has been direct and intentional or indirect and inadvertent, the exclusion of some from the benefits of research violates the societal commitment to justice. A commitment to distributive justice in research imposes obligations on, and concerted activities by, all partners in the research ethics enterprise—researchers and participants, sponsors and institutions, REBs and the public. All have important roles to play in ensuring a fairer distribution of the benefits and burdens of research, as part of a just research agenda. As the following articles make clear, distributive justice imposes on researchers and REBs a duty not to discriminate. Sometimes it may impose positive duties to conduct or include disadvantaged groups in research involving humans. (P. 55)

In the revised text, recognition of the need to involve women and other systematically disadvantaged groups in setting the research agenda has been obscured. Furthermore, the text conflates the problem of exclusion from research participation with the problem of exclusion from setting the agenda. These two distinct types of exclusion must be addressed separately to highlight specific concerns about setting the research agenda and to recognize that different groups (e.g., sponsors, funding councils, REBs, researchers) have different responsibilities with respect to the two sorts of exclusion.

Final Reflections on Process

At this point in the process, we ask ourselves why we have had such limited success in having our concerns fully translated into the final document released by three councils as the *Tri-Council Policy Statement*. Our general conclusion is that we made the mistake of restricting our efforts to offering carefully structured moral arguments (and expecting comparable moral arguments to counterclaims that were to be rejected) in a situation that demanded more explicitly political action. On the basis of our experience, three lessons can be learned.

First, it is important to periodically reevaluate one's overall political strategy so as not to miss an opportunity to pursue other, possibly more effective, strategies. In this instance, for example, we might have shifted from our strategy of attempted persuasion to one of attempted circumvention. In retrospect, we could have gone beyond lobbying the Working Group to actively lobbying the sponsoring councils. We might also have sought to work more closely with the federal government's Women's Health Bureau, since the three funding councils receive their money from the federal government.

Second, when choosing a strategy for political action, it is important to remember that even apparently powerful bodies have limited options and limited power. We learned from our ongoing interactions with the Working Group that one must be mindful of the constraints under which all parties are operating. As noted previously, the Working Group had to report to three granting councils who retained the authority to accept or reject the new guidelines. Arguably, this placed the Working Group in a difficult position. It had to ensure that the final document would be "acceptable" to the councils. It could not stray too far from the interests of the researchers and expect the councils to endorse the final text. As it happened, the Working Group must have strayed too far since the granting councils exercised the authority they had retained, and after the so-called final version of the revised guidelines was submitted, both the process and the substance were changed. Some of the substantive changes have already been alluded to here, and the "open" process was subverted.

Third, it is important to directly challenge those who resist change. We should have taken more seriously the active resistance of large numbers of members of the research community to being required to change the way they conduct their research activities to better accord with feminist goals. The ethics agenda, like empirical research agendas, is subject to political influence, and, as feminists, we need to become more adept at ensuring it is responsive to our input.

Our dealings with the Working Group helped illuminate many of these theoretical issues for us. We hope that others can learn from our experience and our reflections upon our experience and that we can all move toward increased efficacy in our efforts to make the public policy-making process more responsive to the needs and interests of women and other oppressed groups.

Notes

Françoise Baylis is supported by grants from Associated Medical Services Inc., Toronto, and the Dean's Development Fund, Dalhousie University, Faculty of Medicine.

1. See F. Baylis, J. Downie, and S. Sherwin, "Reframing Research Involving Humans," in Feminist Health Care Ethics Research Network, *The Politics of Women's Health: Exploring Agency and Autonomy*, Susan Sherwin, ed. (forthcoming).

2. Where appropriate, we use the expressions "subject-participant" and "researcher-participant" instead of the traditional terms "subject" and "researcher." We do this, despite the somewhat unwieldy nature of these expressions, because of negative connotations associated with the term "subject," which implies passivity, and positive connotations associated with the recognition that *both* researchers and subjects are participants in the research endeavor.

3. In an effort to correct this omission, the Drugs Directorate of Canada recently issued a policy statement regarding the inclusion of women in clinical trials during drug development. This policy requires "the enrollment of a representative number of women into clinical trials for those drugs that are intended to be used specifically in women or in populations that are expected to include women" (Drugs Directorate, *Inclusion of Women in Clinical Trials During Pregnancy* [Ottawa: Health Canada, September 25, 1996], 2pp.).

4. National Forum on Health, "An Overview of Women's Health," in *Canada Health Action: Building on the Legacy, Volume II: Synthesis Reports and Issues Papers* (Ottawa: Minister of Public Works and Government Services, 1997), 17.

Index

Aberg, Anders, 160, 161
abortion: in China, 143, 145, 147; for
  genetic anomalies, 10, 57, 86, 88–
  90, 99–101, 145, 161–62, 164, 167,
  168, 171n4; late-term, 164–65, 167,
  171n7, 172n14; in the Philippines,
  forced, 229; rates, 88, 99–100, 143;
  right to, 60n12, 74, 125–26, 133,
  134, 164, 171n7–8, 233n8, 246; sex-
  selective, 167, 173n17. *See also* fetal
  reduction
abstraction, 33, 53
abuse: fetal, 36–38; human subjects,
  65, 67, 68, 76, 166; institutional, 7,
  248–49
access: to health care, 48, 67, 71–72; of
  lesbians, to assisted reproduction,
  103, 105–6, 109–10, 112, 114–15,
  117, 118n7
acute radiation syndrome (ARS), 91,
  93, 94, 99
Addelson, Kathryn Pyne, 178
adoption: and contract pregnancy,
  129–30; by lesbians, 104, 106, 117,
  117n5, 118n6
adultery blame, 228, 235n25
Africa, West, 225–26, 229–30, 234n15
*African-American Perspectives on Bio-
  medical Ethics*, 75
African Americans, 37, 53, 56, 65, 67–
  68, 71, 72, 180–81

aging, 143, 146, 149, 203–4, 209–10,
  212
AIDS, 215n5, 227. *See also* HIV
Alzheimer's disease, 20, 188
American Medical Association, 68, 72
American Society for Reproductive
  Medicine (American Fertility Soci-
  ety), 51
Americans with Disabilities Act,
  60n9, 178, 185–86, 189–90, 194
amniocentesis, 88
Amundson, Ron, 187
Amy (in Heinz dilemma), 56, 57,
  60n13, 61n14
anencephaly, 50, 57, 96
Annas, George, 49, 128
Archives of Internal Medicine, 108
Arneson, Richard, 131, 137n4
Asch, Adrienne, 77, 193
Asia, Southeast, 225, 229–30
Atomic Energy Agency, International
  (IAEA), 90, 100
Australia, HRT and menopause in,
  208–9, 214, 215n1, 216n12
Australian Menopause Society, 209
autokoenomy, 41
autonomy: of cocaine-using pregnant
  woman, 39, 42; in contract preg-
  nancy, 10, 128–31, 134, 136–37;
  limited, 73, 74; as more options,
  196–97, 246; as principle, 53–54,

269

56, 60n12, 68, 70, 72, 244; reproductive, 112, 121–25, 128, 131–32, 163
Avis, N.E., 208

Baby K, 50–51, 56–57, 60n9, 61n14
Baier, Annette, 25, 29
"A Balanced Approach to the Menopause," 207–8
Barbre, J.W., 206
Barron, Karin, 182
Bartky, Sandra, 22
Baylis, Françoise, 12, 258, 260, 264–65
Beauchamp, Thomas, 49, 60n11, 247
Beijing, 142, 145, 146, 151, 155n1, 224
Bekker, Corinne, vii
Belarus, 86, 90, 91, 94, 95, 97, 98, 101n1; Institute for Hereditary Diseases, 98; National Academy of Sciences, 98; School of Medicine, 98
Bender, Leslie, 243
beneficence, 53, 56, 60n12, 70
Benhabib, Seyla, 75
bioethics: class and, 67, 76n1; deep structure of, 53, 57, 69–70, 74, 76; and "difference," 65–76; elitism in, 46–51, 61n15, 71, 74 ethnicity and, 8, 65–76; feminist. *See* feminist bioethics; gender and, 48–49, 65–77; history of, 7, 46, 65–66, 68–69, 75, 159; mainline/Western, 8, 33, 53, 73, 74, 154, 247, 249; race and, 8, 65–77; strategies for change in, 57–59, 75–76, 241–49; texts, 6–8, 49, 56, 68, 75, 77n2, 247
biotechnology, 243
birthrate, in former USSR, 87–88, 89, 99
birthweight, 96, 166, 172n15
"Black White Disparities in Health Care," 68
blame-the-victim, 52, 54, 55, 143, 185, 228, 230
blameworthiness, 36, 40, 52, 54, 168, 185, 228, 230
blindness, 182, 186, 187, 188, 189, 192
bodily integrity, 121, 124–27, 135

Boetzkes, Elisabeth, 10
Bordo, Susan, 178
British Medical Association, 107
British Parliament, 183
Brock, Dan, 188
Brody, Howard, 241
Brown, Louise, 7, 160
Brussels, donor insemination for lesbians, 109–10, 112–13
Buchanan, Allen, 192–94, 199
Bundo all-women secret society, 237n45
Bunkle, Phillida, 240
Brody, Howard, 241
Calhoun, Cheshire, 23
Callahan, Joan, viii, 164, 167, 172n14, 215n4
Canada: contract pregnancy in, 130, 136–37; Medical Research Council, 253; National Forum on Health, 265; Natural Science and Engineering Council, 254; Reform Party, 132; Research Ethics Boards, directives to, 256–60, 263, 266; Royal Commission on New Reproductive Technologies, 132; sex selection in, 132–34, 137; Social Sciences and Humanities Research Council, 253–54; Supreme Court, 135–36; Women's Health Bureau, 267. *See also* Feminist Health Care Ethics Research Network; Tri-Council Working Group on Research Involving Humans
cancer, 70, 97, 188; cervical, 216n9, 240; leukemia, 91–92; thyroid, 90, 92–94, 97, 99
Caplan, Arthur, 55–56, 60n10
Card, Claudia, 26
cardiac puncture, 160, 164, 171n4
care, ethic of, 9, 17, 22, 26, 27–29, 33–35, 73, 244, 247
caregiver–recipient dyad, 20
caregivers, 17, 29, 39–41, 49, 182
care-receiver, 39–40, 41
caretaking, 22, 24, 29, 30n6, 50, 188

Caring: A Feminine Approach to Ethics and Moral Education, 39

Carlson, Rick, 54

Carse, Alisa, 9, 33–34

casuistry, 72, 74

cesarean section, 50, 57, 61n14, 70, 74; after IVF, 153

Charlesworth, Hilary, 233n9, 234n19

Chernobyl (nuclear power plant explosion), 11, 85–86, 88–101; cancer after, 90–94, 97, 99 clean-up workers (liquidators), 91, 92, 94; cover-up, 89, 93, 99; genetic risk from, 85, 88–91, 94–96, 99–101; psychological effects after, 92–94, 96, 99

Chervenak, Frank, 163, 172n12

chief executive officer (CEO), 45, 52, 53, 59, 59n2

child: commodification of, 128–29; disabled, 144–45, 161; and health insurance, 57–58; pressure to have, in China, 144–45, 149, 152 superior, 142, 145, 146, 154, 168, 197;

childbirth: cultural beliefs, 12, 69, 227–28, 229, 235n25; mortality in, 12, 225, 227–29, 233n11–12, 234n16, 235n23, 236n34–36; in the Philippines, 229, 236n34–36; premature, 160, 166, 171n3; in Russia, 88–89, 99

childlessness, discrimination for, 144

Childress, James, 49, 60n11, 247

China: drugs in, 153–54; infertility in, 12, 141–44, 146–55; modernization, 12, 146, 148–49, 150–51; rural issues, 144–45; technology emphasis, 141, 148, 149

Chinese Communist Party, 143, 144

chlamydia, 147

choice, 4, 244, 246

chorion biopsy, 88

chorion villus sampling, 88, 167

circumcision, female. *See* FGS

civil rights movement, 7, 66, 185

class issues, 67, 76n1, 242, 243

cloning, 10, 55, 57, 105

cocaine. *See* pregnant women, cocaine-using

Code of Ethical Conduct for Research Involving Humans, (Canada), 255–61, 263

Cole, Helene, 38

colonialism, 12, 226, 235n20

community as value, 70, 239, 244–47

Coney, Sandra, 240

confidentiality, 74, 108, 114–15

connectedness, 35, 190, 194

context, 53, 70, 72, 75, 163, 182, 235n20

contraceptives, 89, 213; in China, 143–44, 148; in former USSR, 89; required for subject-participants, 254, 260–61

contractarianism, 18, 19, 25, 194

contract pregnancy: adoption analogy, 129–30; autonomy and, 10, 128–31, 134, 136, 246; state regulation of, 129–30, 134, 136–37;

Cook, Rebecca, 161

Cornell, Drucilla, 10, 121–22, 125–28, 133–36, 137n2–3

Court of Appeals, Fourth Circuit, 50

Crenshaw, Kimberle, 75, 180–81

crisis orientation, 55–56, 58

cross-cultural perspectives, 12, 142, 154, 234n18, 235n20

Crow, Liz, 186

cultural practices, 224, 227–28, 229, 230–31, 235n25

cultural relativism, 224, 233n9, 234n19, 235n20

cultural rituals, 231, 237n45–46

cultural values, 70, 223, 231–32, 234n18, 236n41, 237n43

danwei, 143, 149, 151, 152

Davis, Dena, 195–99

Deaf community, 195–96, 198; loss of children, 198; parenting ability, 196–99;

deafness, 187, 192; as disability, 195–99; social model of, 195

deductivism, 72

degradation prohibition, 127–28, 134–35

democratic political morality, 192
deontologists, 36–37, 39
dependency, 24, 25–26, 30n6, 41, 194, 244
diabetes, 188, 204, 209, 216n8
dichotomies, 3, 56
Dickson, G. L., 212, 216n10
difference as issue for bioethics, 65–76, 243, 245
Dillon, Robin, 20, 21
"Disability, Handicap, and the Environment," 187
disability: activists, 167–69, 173n17, 178, 182, 185, 188–89, 192, 199; Buchanan's scheme for, 192–94; conceptualization of, 168, 183–84, 187, 189–90; discrimination, 6, 177, 183, 189, 190, 194, 199; medical model of, 184–85, 190, 194; menstruation as, 183; mental. *See* mental disability; as moral weakness, 52, 185; perspective, 10, 178, 182, 186, 190, 198–99; from prematurity, 171n3; social model of, 10, 178, 185–86, 190–94, 199; studies, 178, 186 *See also* abortion; child, disabled; deafness
disabled persons, 28, 51–52, 55, 58; ability to parent, 181, 182, 195–99; sexual needs, 184
discrimination: antilesbian, 104–6, 109, 116–17, 118n7; for disability, 177–78, 189–91; racial, 27, 65–67, 135
disease: definition of, 204–7, 209–11, 214, 215n7, 216n9; vs. disability, 186–89; menopause as, 203–4, 207–12, 214, 215n1; *See also* estrogen deficiency disease
disempowerment, 9, 10, 22, 42, 49, 52, 184, 199
doctor-patient relationship, 65, 67, 68, 70
dominant cooperative scheme (Buchanan), 192–94
dominant discourse, 241, 244–45, 247

domination, 25, 33, 34, 46
Donchin, Anne, vii, viii
donor insemination, 103, 105–11, 113–15; in Brussels, 109–10, 112–13;
Downie, Jocelyn, 12
Down's syndrome, 95–96, 168–69, 171n5
Dresser, Rebecca, 47, 49, 59, 60n7
drug trials, 69, 256, 261–62, 268n3
dualisms, 56, 243

economic insecurity, in former USSR, 87, 89, 93, 96, 98
economic issues, 29, 46, 52–53, 58, 59
Edwards, Robert G., 160
Effah, Josephine, 235n22
eggs, human, 55, 105, 118n9
Elliott, Carl, 60n5
embodiment, 121, 123–24, 129, 244
Emergency Medical Treatment and Active Labor Act, 60n9
empathy, 21, 23
empowerment of women, 12, 225, 226, 231–32
endometriosis, 152
Engelhardt Jr, H. Tristram, 7, 244–45, 250n7
Englert, Y., 110
entitlements, 189, 196
environment: built, 178, 185, 186, 188, 190, 192; effects from Chernobyl on, 90, 91
episiotomy, 228
equality, 10, 42, 46; of women, 121–22, 125, 127–28, 131–33, 135–37, 179
essentialism, 75, 207, 233n9, 250n3
estrogen deficiency disease, 203, 209–11, 214, 215n1
estrogen levels, 203, 205, 207–10
Ethical, Legal, and Social Implications Branch, 55
ethic of care. *See* care, ethic of
ethic of justice. *See* justice, in ethics
ethics: committees, 46–47, 51, 60n9–10; communicative, 75; consultants, 46, 47, 51, 52, 60n10; femi-

nist, 17, 177, 182, 253, 267. *See also* bioethics; feminist bioethics
ethnicity and bioethics, 8, 65–76
eugenics, 65, 67, 141, 145–46, 148, 168
Europe, Eastern, 85, 89, 98
European Community (EC), 90, 100
euthanasia, 243
Evans, Mark, 164, 171n8
exclusion: for disability, 190–194; of ethnic subject-participants, 255, 258, 259–60, 262–64; of women subject-participants, 254, 256–59, 262
exploitation, 17, 19, 21, 29, 76, 246, 247, 256
Extraordinary Bodies, 202n57

FAB. *See* Feminist Approaches to Bioethics
family (birth) planning, 143–45, 147, 224, 235n21
Farrell Smith, Janet, 75
Feinberg, Joel, 135
female, devaluation of, 132–33, 136, 167, 183–84
female body, surveillance of, 143–44, 145
female genital surgery (FGS). *See* FGS
Feminine and Feminist Ethics, 233n13
femininity, 29, 34, 126–27, 130, 134, 136, 205, 207, 210
feminism: core, 2–4, 223; and disability perspective, 177–80, 182, 198–99; policy making, 8, 12–13, 57–59, 253–54, 267–68; postmodern, 234n13, 245, 250n8; schools of thought, 129, 234n13; Western, 224, 226, 229–30, 235n20. *See also* feminist bioethics; ethics, feminist
Feminist Approaches to Bioethics (FAB), International Network on, vii, 49, 61n15
feminist bioethics: action plans, 57–59, 240, 242–49 core themes, 2, 8, 55, 75, 163, 170, 172n12, 223, 243, 245–47; published work, vii, 1, 59, 72, 75, 159, 243; reproductive is-

sues, 121, 154, 163–65, 169, 171n6, 173n16,18, 244. *See also* feminism and disability perspective
Feminist Health Care Ethics Research Network (Canada), 253–268
Feminist Perspectives in Medical Ethics, 9
fertility: research on, 263; women's value from, 227–29. *See also* infertility
fetal reduction: elective, 11, 159, 160, 162–64, 166–67, 169, 170; informed consent, 165–66, 170, 173n16; methods of, 160, 171n4; multifetal, 160, 161, 162–63, 165, 171n8, 172n9, 173n15; risks from, 165–66, 170, 172n10, 173n15; selective, 11, 160–61, 163, 167, 169, 171n5, 172n13; in twin pregnancy, 159–60, 162, 166, 169
fetus, 5, 163; anomalous, 10, 11–12, 145, 161, 162, 164, 165, 167; cocaine-exposed, 35–39; vanishing, 162, 172n11. *See also* maternal–fetal relations; pregnant women
fetus-as-patient, 172n12
FGS (female genital surgery): eradication of, 229–31, 237n42, 44–46; feminist views on, 230, 235n20, 245; as torture, 230, 236n38, 41, 237n43
First National Exhibition on Eugenics, 145
Fletcher, John, 108, 109, 164
Foucault, Michel, 54, 144, 185
Foundations of Bioethics, 244
Fox, Renee, 50, 53, 56, 60n4,11
fractures, hip. *See* hip fractures
Freedman, Benjamin, 47, 51
freedom, 35, 223, 250n7
Freud, Sigmund, 10
Friedman, Marilyn, 27
Frug, Mary Joe, 74–75
Fruits of Sorrow: Framing Our Attention to Suffering, 191

gamete intrafallopian transfer (GIFT), 105, 141, 148

gap between rich and poor, 9, 45–46,
48, 52, 59, 68, 74, 75
Garcia, Jorge, 75
Gatens, Moira, 10
Gauthier, David, 19
Geller, Gail, 77, 193
gender: analysis, 2, 49, 66, 73, 249,
256; bias, 28, 34, 168, 255, 260; in
bioethics, 53, 65–77; equity, 13, 35,
48–49, 231, 224, 254, 257; role, 40
General Maternal and Child Health
Project, 39
genetic counseling, 88; of deaf parents, 194–96, 198–99
genetic dilemmas *re* disability, 178,
179, 199
genetic disorders, 108, 144, 167, 168;
from Chernobyl, 85, 88–91, 94–96,
99–101
genetic engineering, 193, 243
genetic testing, 67, 86, 88–90, 99–100,
101, 167–69, 188, 243
Georgetown mantra, 49
Gilligan, Carol, 56, 250n6
gishiri cut, 227–28
Glennie, Evelyn, 192
Goldzieher, Joseph, 215n1
Gorowitz, Samuel, 6–7
Grady Memorial Hospital (Atlanta),
Project Prevent, 39
grass roots, 13, 248
Great Britain: disability movement in,
185; Down's syndrome in, 95; thyroid cancer in, 91, 97
Grosz, Elisabeth, 245
groups, significance of, 54, 70, 76, 180,
226
Gunning, Isabelle, 235n20, 236n41

Hahn, Harlan, 190–91
Hall, Alison, 172n13
Hampton, Jean, 19
Handwerker, Lisa, 7, 12
Hanna, William, 181
Harris, John, 171n5
Harris, Leonard, 54
Harrison, Kelsey, 228

Hart, H.L.A., 135
Hastings Center Report, 48, 195
Hauerwas, Stanley, 250n7
health: reproductive, 223–28, 231,
232n5, 235n21; sexual, 224, 225,
232, 233n5; as virtue, 52, 55
health care: children and, 57–58; in
China, 142, 148–49, 153, 154; differential treatment, 1–2, 67, 72, 74,
76; in U.S., 9, 45, 48, 51, 74
health care workers, 45, 48–49, 52–53,
55, 59, 69
health insurance: for children, 57–58;
infertility coverage, 117n4; lack of,
1, 48, 69, 74, 76n1, 188, 189; for lesbian insemination, 103, 105, 106,
117
health maintenance organizations, 1,
45, 52
heart disease, 208, 212, 213, 216n8
Hecht, David, 237n45
Hegel, G. W. F., 185
Heinz dilemma, 56, 60n13. *See also*
Amy; Baby K
Held, Virginia, 24
Heumann, Judy, 182
hierarchy, 8, 9, 33, 127, 248
hip fractures, 203, 210–14, 215n2,
216n11–12
HIV, 89, 113, 118n7, 215n5, 227
Hoagland, Sarah Lucia, 30n6, 41
Holmes, Helen Bequaert, vii, viii, 1, 7,
9, 12
homophobia, 6, 26, 240, 247
homosexuals, 7, 52. *See also* lesbians
Hooker, Michael, 234n14
hormone replacement therapy. *See*
HRT
hospitals, 45, 56, 59; addicted babies
in, 35–36, 39; length of stay, 45, 53,
59n1
hot flashes, 207, 208, 211
housekeeping issues, 55–56, 57, 58
HRT (hormone replacement therapy):
promotion and cost-benefit claims,
203, 212–14, 215n1–2, 216n15; risk
reduction, 211–14, 216n12

Human Genome Diversity Project, 55, 243
Human Genome Initiative, 67, 193
human subject research. *See* research with human subjects
Hunt, Mary, 58
Hurler's syndrome, 171n5
hysterosalpingogram, 147, 150

Illich, Ivan, 54
illness compared to disability, 186–89
inclusion, 74, 178–82, 194
individualism, 53–54, 70, 197
individuation, 121, 125–28, 131, 132
inductivism, 72, 74
inequality, 3, 20, 24–26, 49, 179, 183, 247, 250n5. *See also* equality
infant mortality, 225, 227, 229, 233n10
infants, cocaine-exposed, 35–39
infertility: in China, 12, 141–44, 147–54, 155n1; clinics, 105, 114, 115, 137, 141, 147–48, 150–51, 154, 155n1; diagnosis of, 98, 104–7, 110–17; industry, 12, 141, 146–53, 170; relational/temporary, in lesbians, 11, 103–117, 118n10; treatment of, 107–8, 151–53, 160, 170. *See also* in vitro fertilization
infibulation, 236n38
informed consent, 55, 65, 137, 163, 165, 166, 173n16
Inlander, Charles, 54
insemination: as industry, 114; physician-assisted, 107–8, 112, 115–16, 117n3; self, 104–5, 114, 115, 117n3, 118n7
insurance. *See* health insurance
Inter-African Committee on Traditional Practices, 235n24, 236n30, 237n44
interdependence, 21, 25, 26, 28, 194, 196–97
interdisciplinarity, 243, 244, 253
International Association of Bioethics (IAB), vii
International Conference on Popula-

tion and Development (ICPD)(-Cairo), 224, 232n5, 233n8, 236n33
International Planned Parenthood Federation Charter, 233n7
in vitro fertilization (IVF), 105, 111, 117n4, 118n9, 134, 136, 137, 141–42, 146–48, 151, 153–54, 155n5

Jaggar, Alison, 34, 177, 184
Jameton, Andrew, 6
Japan: menopause symptoms in, 208; triplet births in, 171n2
Jones, Nikki, 12, 13
jurisprudence, 76, 136. *See also* legal remedies
justice: in ethics, 33 36, 38–39, 42, 76, 177, 223, 242; in families, 18–20, 29n1; as principle, 53, 61n13, 70, 74, 232n3; social, 3, 27, 193, 224, 257, 266

Kallianes, Virginia, 177
Kant, Immanuel, 6–7, 19, 70, 137n3
Kegley, Jacquelyn Ann, viii
Kennedy Institute of Ethics, vii
Kennedy Institute of Ethics Journal, 50
Kent, Deborah, 182
Kenya, 237n46
Kiragu, Karungari, 237n42–43
Kleinman, A.M., 234n18
Kohlberg, Lawrence, 56
Komesaroff, P.A., 215n4
Koso-Thomas, Olayinka, 237n45

labor, obstructed, 227–29, 235n23,25. *See also* childbirth
Ladd, Rosalind Ekman, viii
Lamb, Brian, 184
Langton, Rae, 135
Layzell, Susan, 184
legal remedies, 56, 60n9, 127, 128, 135, 137. *See also* jurisprudence
lesbian baby boom, 113
lesbians: as infertile, 3, 10–11, 104, 105, 109–10; as parents, 103, 106–7, 110–13, 117
Little, Margaret, vii, 30n8

Logothetis, Mary Lou, 208
Lukins, Sheila, 177

Mabry, Marcus, 237n45
Macedo, Donaldo, 250n2
Mackenzie, Catriona, 121–25, 128, 129–30
MacKinnon, Catharine, 73, 135
Macklin, Ruth, 223, 232n3–4
Maendeleo Ya Wanawake, 237n46
Mahowald, Mary, viii
male-as-norm, 33–34, 69, 205, 244, 248, 255–56
mammography, 212
managed care, 1, 45–46, 58
marginalization, 52–57, 59, 66, 71–75, 241, 248; of the disabled, 1, 10, 52, 177, 182, 188; in health care, 9, 52, 56, 249; of women, 12, 66, 184, 239; of women of color, 1, 53, 66, 180–81
margin-to-center, 66, 239–40, 247–49
market economy, 11, 24, 45, 149–50
marriage: child, 227, 228, 235n29; and disabled women, 181
Martha's Vineyard, 195, 201n53
Martin, Emily, 216n10
Marx, Karl, 185
Marxist feminists, 129
Massachusetts Women's Health Study, 208
masturbation as disease, 206
maternal–fetal relations, 35–40, 42
maternalism, 24, 40, 173n18
McCullough, Laurence, 163, 172n12
McRea, F. B., 215n1
medicalization, 60n7; of disability, 179, 184–85, 190; of donor insemination, 114–15; of women, 69, 183–84. *See also* menopause
Medical Research Council of Canada, 253
medical schools, bioethics in, 46, 69, 241
medicine: cognitive authority, 46–48, 51, 54, 60n6–7; as commodity, 150, 153; power of, 51, 58, 114, 170, 203,

241, 242; sexism in, 48, 49, 65, 71, 240; social authority, 45–48, 60n6, 203; as value-free, 204, 215n3; Western, 46, 215n4
Mendelsohn, Robert, 54
menopause: Australian Society, 209; conceptualizations of, 204, 207–8, 210, 212, 214, 215n4, 216n10; as disease, 11, 69, 203–4, 207–12, 214, 215n1; nineteenth-century, 206, 216n8
menstruation, as disability, 69, 183
mental disability, 35, 145, 161, 178, 182–83, 189
Meyers, Diana, 123–24, 132, 197
Miles, Steven, 48
Mill, John Stuart, 7, 70
Ministry of Public Health, China, 150
minorities, inclusion in research, 72, 204, 255, 258. *See also* African Americans; disabled persons; lesbians
Minow, Martha, 177, 185
miscarriage, 98, 100, 101, 143, 162, 166
Mitchell, Juliet, 234n13
moral agency, 121, 122, 124, 250n10
moral harm, 195
moral insanity, 206
Moral Problems in Medicine, 6–7
moral standards, community, 246
mortality: infant, 225, 227, 229, 233n10; maternal, 12, 225, 227–29, 233n11–12, 234n16, 235n23, 236n34–36
mother–child dyad, 17, 24–25, 30n6, 40
Mother Father Deaf: Living Between Sound and Silence, 198
motherhood: chosen by single women, 103, 105, 116; compulsory, 113, 117, 142, 144, 145, 149; contract. *See* contract pregnancy; as gender norm, 113, 152, 154, 182, 184; social meanings of, 10, 122–25, 128–30; surrogate. *See* contract pregnancy

Mullett, Sheila, 40
multiple births, 153, 160, 161
Murphy, Julien, 10–11
Murphy, Margaret, 228, 235n29
Murphy, Timothy, 117n2

narrative ethics, 72, 247
National Defense Scientific Working
    Committee (China), 145
National Forum on Health (Canada),
    265
National Task Force on VVF (Nige-
    ria), 229, 235n22, 236n33
Native Americans, 55, 72
Natural Science and Engineering
    Council (Canada), 254
Nazi concentration camps, 8, 22, 65,
    67
Nelson, Hilde Lindemann, 9, 48, 193,
    194
Nelson, James Lindemann, 48, 193,
    194
neonates: cocaine-exposed, 35–36, 38,
    40 death of, 57, 171n3; intensive
    care for, 50, 55, 58
Network. *See* Feminist Approaches to
    Bioethics; Feminist Health Care
    Ethics Research Network
New Zealand: cervical cancer scan-
    dal, 7, 240–41; Code of Rights, 241;
    Committee of Inquiry, 240
Nicholas, Barbara, viii, 7, 12
Nigeria, 225–31, 234n15–16, 235n21–
    22,24–29, 236n30–32,34, 237n44
Noddings, Nel, 21, 24, 39
nonmaleficence, 53, 70
Noonan, John, 7
normal, definition of, 144, 204–5
norms. *See* roles
Norway, 88, 96, 208
nuclear power risks, 11, 85, 90, 99. *See
    also* Chernobyl
Nuremberg, 67
nurses, 52, 116, 236n31, 237n44; as
    caretakers, 22, 23–24, 28
nursing theory, 250n6
nurturing role, 28, 40, 181

offense principle, 135
old age care, 143, 146, 149
Oliver, Mike, 178, 199
one-caring, 39, 40
one-child policy, 143, 145
"One Decade After Chernobyl" con-
    ference, 85, 90–94, 96–101
Ontario Law Reform Commission,
    130
oppressed groups: learning from,
    5–6, 8; as researchers/ and deter-
    mining priorities, 262–63; as sub-
    ject-participants, 255
oppression: of disabled persons, 179,
    180, 184, 189, 199; eradication of,
    225, 263; as ethically wrong, 2, 46,
    223, 245–47; expectation to repro-
    duce as, 113, 117, 144–45, 149, 152;
    history of, 183–184, 186; of women,
    2, 6, 28, 177–78, 180, 183–84, 189,
    191, 197–98, 223, 239
organ transplants, 50, 55, 57, 70, 241,
    243
osteoporosis: Australian National
    Consensus Conference, 216n12;
    costs of, 203, 211, 213–14, 215n1–2;
    definition of, 204, 216n11; fractures
    from, 203, 210–14, 215n2,
    216n11–12; HRT for, 203, 208; risk
    reduction, 208, 210–14, 216n12
Other, symbolic, 127–28
Ott Institute of Obstetrics and Gyne-
    cology (St. Petersburg), 88–89
Our Bodies, Ourselves, 7
ovarian hyperstimulation, 105
Overall, Christine, 128, 164, 167,
    172n6, 173n18

pain, 186, 187, 189. *See also* suffering
paraplegia, 187, 188
parenting, 112, 122, 124; in Deaf com-
    munity, 195–99; and lesbian
    rights, 103, 106, 117
particularity, 8, 18, 33, 163
paternalism, 7, 11, 21, 56, 60n12,
    173n18
paternity, 106, 114–15

patients: experiences of, 52, 55–56, 206–7; generic, 65, 68; interests and rights of, 45, 47, 49, 51, 66, 75; silencing of, 205–6
patriarchal discourse, 239, 240, 242, 243, 247, 248
personhood, 125–27, 135
philanthropy: feminist, 224, 226, 231–32; moral role of, 224, 225, 226, 231–32, 234n14
the Philippines, abortion and childbirth, 229, 236n34–36
Philippines Safe Motherhood Survey, 36, 236n34
physicians, 11, 45, 46. *See also* medicine
Physicians for a National Health Program, 58, 61n16
Pittsburgh Protocol, 50
policy making, 12, 35, 59. *See also* feminism, policy making
poor people, 45, 52, 55, 58, 73, 74, 76n1; research participation of, 255, 264. *See also* poverty
population policy, 67, 142–44, 147, 224
pornography, 126, 135
postmodernism, 2, 245, 246, 250n8
potassium chloride injection, 164, 171n4
poverty, 225, 226; in Russia, 87, 93, 96, 99
power analysis, 13, 53, 68, 75, 244, 248, 250n10
power imbalance, 8, 9, 25, 245, 246
power relationships, 12, 46, 49, 56, 163, 226, 234n18, 250n7
pregnancy: continual, 225; cultural beliefs, 227–28, 229, 235n25; early, 227, 228, 235n25, 29; embodiment, 121, 123, 124, 129; loss, 98, 100, 101, 143, 162, 166; meanings of, 121–123, 128, 163, 172n12; mortality in. *See* childbirth, mortality in; multifetal, 160, 161; *See also* contract pregnancy

pregnant women: cocaine-using, 9, 35–42, 74; exclusion from research, 255, 258, 260, 261; as fetal containers, 124, 127; surveillance in China, 145
prematurity, 160, 166, 171n3
Preston, Paul, 198
Principles of Biomedical Ethics, 68
principlism, 6, 18, 49, 53, 60n11–12, 70, 72, 74
privacy, 124, 125
privilege, 5–6, 74, 75, 239, 243, 248–49, 250n5
profits, 49, 59n2; infertility industry, 148; pharmaceutical firms, 213
prostitution, 227
punishment, theories of, 36–38
Purdy, Laura, viii, 1, 2, 59, 137n8

quadruplet pregnancies, 161, 171n2–3

R. v. Butler, 135
R. v. Keegstra, 135
race: and bioethics, 8, 65–77; issues, 243, 244, 246; and research participation, 255, 258, 259–60, 263–64
race theory, critical, 54, 76
racism, 6, 26, 37, 71, 130, 135, 240, 247
radiation: exposure to, 11, 85, 90–93, 97, 99–101; genetic anomalies from, 85, 88–90, 94–98, 99–101
radiation sickness. *See* acute radiation syndrome
Rawls, John, 67
Raymond, Janice, 173n16
receptivity, 22, 23
recruitment of subject-participants, 255, 259
reductionism, 60n11
Rehabilitation Act, 60n9
relationships, human, 20–21, 33, 39–40, 53, 244
relativism, 135n20, 224, 233n9, 234n19, 244–46
religious values, 228, 231–32
Reproduction, Ethics, and the Law, 172n14

reproduction: assisted, 10, 103–5, 110, 114–16, 160, 241; autonomy and choices in, 4, 8, 85, 90, 99, 115, 121–25, 128, 131–32; in lesbians. *See* lesbians; as norm, 141–42

reproductive rights, 104–5, 112, 171n6, 224, 233n6–7

reproductive technologies, new (NRTs), 141–42, 146, 154, 166. *See also* contract pregnancy; in vitro fertilization; sex selection

Research Ethics Boards (Canada), directives to, 256–60, 263, 266

research with human subjects: ethics, 7, 246, 253; exploitation in, 65, 76; setting priorities, 263–66; women as subject-participants, 65, 72, 74, 254–60

resource access, 154, 223, 226

retention of research subjects, 255, 259

rich-poor gap, 45–46, 48, 52, 59; in China, 150–51

rights, 54, 60n12, 76. *See also* reproductive rights

Roberts, Dorothy, 73–74

Roe v. Wade, 7, 164

Rogers, Wendy, 11, 216n13

Rogovsky, Elizabeth, 181

roles. *See* femininity; male-as-norm; motherhood as gender norm; nurturing role; stereotyping of women

Roosevelt, Franklin D., 192

Rorty, Mary, V., 11

Royal Commission on New Reproductive Technologies (Canada), 132

Rubenfeld, Phyllis, 177

Ruddick, Sara, 24, 40 Russia, 89, 91, 94, 99, 101n1. *See also* Chernobyl

Rust v. Sullivan, 74

Safe Motherhood Initiative, 225, 233n12

safe sex, 232n5

Sambo, Hajiya Amina, 236n33

Sanders, Cheryl, 56

San Francisco conference (FAB), vii–viii

Sankar, Pamela, 55

Sartre, Jean-Paul, 7

scandals, starting bioethics, 7, 65, 66, 68, 75, 76, 240–41

Schweickart, Patrocinio, 29

Self, Society, and Personal Choice, 123

self-definition, 121–24, 130, 132

self-determination, 54, 121–25, 129–30, 133, 136

self-insemination, 104–5, 114–15, 117n3, 118n7

self-interest, 19, 29n2

self-respect, 24, 127, 130, 132, 136, 188

self-sacrifice, 17–22

self-understanding, 122–24, 127–29, 131

sexism. *See* medicine, sexism in

sex selection: legal remedies, 10, 134–37; abortion for, 133–34, 167, 173n17; in Canada, 132–34, 137; futuristic claims, 131–32; preconception, 134, 136, 137; symbolism of, 128, 132–33, 167, 196–97 sexual difference, 127, 133, 134

sexual health, 224, 225, 232, 233n5

sexuality, 108, 126, 224, 230, 234n13

sexually transmitted diseases, 147, 233n5

Shakespeare, Tom, 182

Shanner, Laura, viii, 7, 11

Sherwin, Susan, vii, viii, 2, 6, 12, 177, 182, 183, 215n4, 223, 243, 245–46

sick persons, 51–52, 187

Sierra Leone, 237n45

Silvers, Anita, 10, 173n17

Sims, Marion, 68

social change, 240, 247–48, 249

social justice, 26–28, 249

Social Sciences and Humanities Research Council of Canada, 253–54

Somalia, 230

Soviet Union. *See* USSR

specialization, resistance to, 242–43, 247, 250n2

Spelman, Elizabeth, 178, 191
sperm banks, 104–5, 114, 115
Stangl, Teresa and Fritz, 22
Steptoe, Patrick, 160
stereotyping, 76; of the disabled, 184; racial, 71, 135; of women, 10, 17–18, 73, 122, 130–31, 184
sterility, 111
sterilization: abuse, 7, 67; of the disabled, 145, 180, 182; policy, 145, 180; reversing of, 148
stigmatization, 67, 71, 185; of multiple children, 153
strategies: political, 10, 13, 247–49, 265, 267; theoretical, 242–47
subgroup analysis, 256–59
subject-participant, as term, 268n2. *See also* research with human subjects
subordination of women, 17, 34, 52–53, 73
success rates for IVF, 165
suffering: of daughters, 197–98; of the disabled, 161, 169, 171n5, 180, 186, 190–91; of women, 177–78
surgery, 46, 50, 52, 68, 70
surrogate motherhood. *See* contract pregnancy
Swazey, Judith, 53, 56, 60n4,11
Szasz, Thomas, 54

technology, Western, 7, 12, 45, 141–42, 144, 149, 150
test-tube baby: first, 7, 160; first Chinese, 141, 146, 151, 155n1
Thomson, Rosemarie Garland, 201n57 thyroid cancers, from Chernobyl, 90, 92–94, 97, 99
Tong, Rosemarie, viii, 9, 23, 60n11, 129, 130, 177, 233n13, 243
Tonti-Filippini, Nicholas, 47
Tooley, Michael, 7
transplants, organ, 50, 55, 57, 70, 241, 243
Tri-Council Policy Statement on Ethical Conduct for Research Involving Humans, 259–60, 261–62, 266–67

Tri-Council Working Group on Research Involving Humans (Canada), 253–55, 257–58, 260–61, 267
triplet pregnancies, 161, 171n2–3, 172n11
trisomies (13,18,21), 95–96, 168–69, 171n5
trust, 26, 28
truth-telling, 67, 71, 74
Tuskegee syphilis study, 8, 65, 67–68
twin: pregnancy, 159–160, 162, 166, 169, 171n2–3, 172n15; reduced to singleton, 159–60, 162, 166, 169

U.S. Congress, 57, 59n1, 194. *See also* Americans with Disabilities Act
U.S. Supreme Court, 74
Ukraine, 85, 86, 90, 91, 94, 101n1. *See also* Chernobyl
ultrasound, 136, 137
UNICEF, 235n23, 26
United Nations: Development Program, 236n34; International Conference on Population and Development (ICPD) (Cairo), 224, 232n5, 233n8; World Conference on Women (Beijing), 181, 224, 233n6
universality, 33, 53, 60n11, 70
USSR (former): abortion in, 88, 89, 99–100; health in, 87, 93; instability in, 87, 89, 92–93, 96, 98–99. *See also* Chernobyl
utilitarianism, 38–39, 60n13, 223

Vermeij, Geerat, 192
vesico vaginal fistula (VVF), 227–29, 235n22, 236n33
Voda, Ann, 215n1
vulnerability, 7, 19, 21, 25, 26, 28, 48, 49, 68, 150, 241

Walker, Alice, 230
Walker, Margaret, 18, 60n11
Wapner, R., 163
Ward, Ann, 234n16, 235n27
Warren, Mary Anne, 173n18
Warren, Virginia, 55–56

well-being, 35, 47, 145, 188, 194, 196, 197, 231, 232
Wendell, Susan, 52, 60n6, 188–189
wheelchair mobility, 182, 185, 186, 188
Wikler, Dan and Nancy, 114–115
Wilson, R.A., 205, 209, 215n1
Wolf, Susan, 8, 9, 12, 51, 53, 54, 57, 60n9, 178, 179–80, 185, 243
womanhood: assumptions about, 203–204; initiation into, 231, 237n46
women: blaming of, 28, 143; commodification of, 128–29; differential treatment, 52–53; in health care, 48, 49, 52–53, 249; status of, 17, 34, 52–53, 226, 232, 239; underrepresentation of, 254–57, 262, 264. *See also* feminist bioethics; oppression; research with human subjects; stereotyping
women's bodies, as abnormal, 212, 214, 216n13

Women's Disabilities Bill, 183
women's health advocacy: in Nigeria, 228–29, 230–32, 235n22, 236n30–33, 237n44–46; in the Philippines, 229; strategies for, 226–29, 231–32, 234n18, 235n20
Women's Health Bureau (Canada), 267
women's health movement, 7, 49, 54, 58, 59
women's movement, 54, 66; and disabled women, 177
Wonderwoman and Superman, 171n5
World Conference on Women (Beijing), 118, 224, 233n6
World Health Organization (WHO), 90, 96, 100, 147, 216n11, 233n11

Young, Iris Marion, 41, 177, 184
Youngner, Stuart, 47

Zita, Jacqueline, 212

About the Editors and Contributors

Françoise Baylis is Associate Professor, Bioethics Education and Research, Faculty of Medicine and Department of Philosophy, Dalhousie University, Halifax, Nova Scotia, Canada. She is editor or coeditor of several books in health care ethics, including *Health Care Ethics in Canada* and *The Health Care Ethics Consultant* and has published a number of essays on women and research.

Elisabeth Boetzkes is Associate Professor of Philosophy at McMaster University where she conducts research in health care ethics, philosophy of law, and philosophy of religion. Her recent publications focus on feminist critiques of current reproductive practices, and she is presently working on a manuscript investigating the legal notion of harm from a feminist point of view.

Alisa L. Carse is Associate Professor of Philosophy and a Senior Research Fellow at the Kennedy Institute of Ethics. Her research and teaching focus on social and political philosophy, ethical theory, moral psychology, and feminist theory. Recent publications center on pornography, feminist pedagogy, particularism and affiliative virtue in ethical theory, and the virtues of care in bioethics.

Anne Donchin is Associate Professor of Philosophy and former director of Women's Studies at Indiana University, Indianapolis. A founder and coordinator of the Network on Feminist Approaches to Bioethics, her publications focus on the intersection of biomedical ethics and feminism, particularly personal autonomy and moral issues raised by innovative reproductive/genetic practices. She is the author of *Procreation, Power, and Personal Autonomy: Feminist Reflections* (forthcoming).

Jocelyn Downie is Director of the Health Law Institute and member of the Faculties of Law and Medicine at Dalhousie University, Halifax, Nova Scotia. She has published articles on feminist health care ethics consultation, feminist explorations of assisted death, and reframing human research. She, Susan Sherwin, and Françoise Baylis are coauthors of the recent release *The Politics of Women's Health: Exploring Agency and Autonomy*.

Lisa Handwerker is visiting scholar at the University of California at Berkeley, Institute for the Study of Social Change, and guest lecturer in the Anthropology Department at San Francisco State University. She has published numerous articles on female infertility in China and currently chairs the Council on Anthropology and Reproduction of the Society for Medical Anthropology. An activist, she sits on the board of the National Women's Health Network and serves as a Berkeley Health Commissioner.

Helen Bequaert (Becky) Holmes is an independent scholar with a Ph.D. in genetics. She currently coordinates the Center for Genetics, Ethics and Women in Amherst, Massachusetts. She was a cofounder and coordinator of FAB and organizer of a workshop funded by the Ethics Branch of the National Center for Human Genome Research. In addition to her many articles on reproductive technology, she has edited several books, including *Issues in Reproductive Technology* and *Feminist Perspectives in Medical Ethics* (with Laura Purdy).

Nikki Jones, a native of the United Kingdom, works for a U.S.-based philanthropic agency responsible for supporting reproductive health programs in Southeast Asia. Her previous involvements include teaching, research, advocacy, and community action around issues of gender and reproductive health and rights. Her academic preparation includes a dissertation on French Catholic and North African Muslim women and degrees in social policy, social work, linguistics, and philosophy.

Julien S. Murphy is Professor of Philosophy at the University of Southern Maine, Portland, where she teaches courses in feminist theory, continental philosophy, and medical ethics. She is also a faculty member in the Women's Studies and Honors programs. She is the author of *The Constructed Body: AIDS, Reproductive Technology and Ethics* and editor of *Feminist Interpretations of Jean-Paul Sartre* (forthcoming).

Hilde Lindemann Nelson is Director of the Center for Applied and Professional Ethics at the University of Tennessee, Knoxville. An editor at the *Hastings Center Report* from 1990 to 1995, she writes on issues in bioethics and feminist theory and is the author (with Jim Nelson) of *The Patient in the Family* and *Alzheimer's: Answers to Hard Questions for Family*. She has edited two anthologies, *Feminism and Families* and *Stories and Their Limits: Narrative Approaches to Bioethics* and coedits Routledge's Reflective Bioethics series.

Barbara Nicholas is a pakeha New Zealander, who recently received her philosophy Ph.D. with a specialization in bioethics and currently teaches at the Bioethics Research Centre, Otago University, New Zealand. Her background includes secondary teaching, parenting, scientific and theological training, and community involvements including a play center, church, school committees, women's refuge, and feminist theology groups.

Laura M. Purdy is Professor of Philosophy: Wells College and University of Toronto and Bioethicist: Toronto Hospital and Toronto Joint Centre for Bioethics. Among her publications are *Feminist Perspectives in Medical Ethics* (coedited with Helen Bequaert Holmes), *In Their Best Interest? The Case Against Equal Rights for Children, Reproducing Persons: Issues in Feminist Bioethics*, and *Violence Against Women: Philosophical Perspectives* (coedited with Stanley French and Wanda Teays).

Wendy A. Rogers, after receiving a medical degree, undertook postgraduate studies in general practice and worked in primary health care for a number of years before returning to earn a philosophy Ph.D. in bioethics at Flinders University School of Medicine in South Australia. Her main research interest is the relationship between bioethical theory and clinical practice in the areas of menopause and primary health care.

Mary V. Rorty, a philosopher by training, is currently Director of Advanced Studies at the Center for Biomedical Ethics, University of Virginia. She teaches philosophy, feminism, and bioethics and writes at the intersection of the three.

Laura Shanner is currently the I'Anson Assistant Professor in the Department of Philosophy and Joint Centre for Bioethics at the University of Toronto. She has just assumed a new position as Associate Professor in the Department of Public Health and the John Dossetor Health Ethics Centre at the University of Alberta. Her main areas of interest are reproductive and genetic technologies, feminist bioethics, and bioethics education.

Susan Sherwin is Professor of Philosophy and Women's Studies at Dalhousie University, Halifax, Nova Scotia, Canada. She is author of *No Longer Patient: Feminist Ethics and Health Care* and coordinates the Canadian Feminist Health Care Ethics Research Network, which has recently completed *The Politics of Women's Health: Exploring Agency and Autonomy* (included are contributions by several authors in this volume).

Anita Silvers, Professor of Philosophy at San Francisco State University, recently coauthored *Disability, Difference, Discrimination* with Mary Mahowald and David Wasserman, which explores how formal, distributive, and feminist approaches to justice address the challenge of equalizing people with disabilities. She has published extensively in the areas of aesthetics, ethics and bioethics, disability studies, and public policy.

Rosemarie Tong is the Distinguished Professor in Health Care Ethics at the University of North Carolina at Charlotte. She is author of *Women, Sex and the Law; Ethics in Policy Analysis; Feminist Thought: A Comprehensive Introduction; Feminine and Feminist Ethics; Controlling Our Reproductive Destiny: A Technological and Philosophical Perspective* (coauthor); and *Feminist Approaches to Bioethics.* Her articles on the ethics of care, ethics committees, health care reform, futility, rationing, assisted suicide, women's health, alternative medicine, abortion, surrogate motherhood, assisted reproduction, and cosmetic surgery regularly appear in scholarly medical and ethics journals.

Susan M. Wolf is Associate Professor of Law and Medicine at the University of Minnesota Law School and a Faculty Member in the University's Center for Bioethics. She edited *Feminism & Bioethics: Beyond Reproduction* and has published numerous articles and book chapters including analyses of gender issues bound up with physician-assisted suicide and geneticism.